SHADOWLANDS

Quest for Mirror Matter
in the Universe

SHADOWLANDS

Quest for Mirror Matter in the Universe

Robert Foot

foot@physics.unimelb.edu.au
School of Physics
University of Melbourne
Victoria 3010 Australia

Shadowlands: Quest for Mirror Matter in the Universe

Copyright © 2002 Robert Foot
All rights reserved.

Universal Publishers / uPUBLISH.com
Parkland, Florida • USA • 2002

ISBN: 1-58112-645-X (paperback)
ISBN: 1-58112-644-1 (ebook)

www.uPUBLISH.com/books/foot.htm

The picture on the cover shows the crab nebula–the remnants of a star which exploded in 1054.This photograph was taken at the European Southern Observatory. (Credit: FORS Team, 8.2-metre VLT ESO).

*A CIP catalog record for this book is
available from the Library of Congress.*

Preface

My purpose in writing this book is two-fold. First, many non-specialists ask me to explain the mirror matter idea and the scientific evidence for it. Second, science is so specialized these days that many people who know a lot about one field often know little about another. Mirror matter, if it exists, would lead to rather important implications for several scientific fields, including: particle physics, astrophysics, cosmology, meteoritics and planetary science. Thus, it seemed to me that an interesting challenge would be to write a book explaining the motivation for mirror matter and its evidence which could usefully serve these two communities (that is, both specialists and non-specialists alike). Such a venture, though, is not without risks of various kinds. Let me state at the outset that the mirror matter idea is not established fact; it is an example of cutting-edge science in progress. It is my hope that people who read this book will be infected by, or at least understand, my enthusiasm for this subject, and why I think it is one of the most interesting questions in science at the moment.

The process of writing this book gave me the opportunity to re-think many of the original arguments. Some 'gaps' in my knowledge were filled in, and a few new directions explored. Some material is therefore completely new, although most of it has appeared in the technical scientific literature previously. I have only cited this scientific literature sparingly, but nevertheless I have endeavoured to properly credit the people responsible for the main original ideas.

It seems only yesterday that I learned as a student that mirror reflection symmetry was not respected by the fundamental interactions of nature. Electrons and other elementary particles are, in a sense, 'left-handed'. Although most scientists have simply come to accept that God is 'left-handed', somehow it always bothered me....

One sunny afternoon in May 1991 a rather remarkable thought occurred to me. While playing with an unrelated idea, it suddenly struck me that there was a subtle yet simple way in which mirror reflection symmetry could still exist. Nature's mirror could be unbroken if each type of ordinary particle has a shadowy mirror partner. The left-handedness of the ordinary particles could then be balanced by the right-handedness of the mirror particles. So there you have it, mirror reflection symmetry can exist but requires something profoundly new. It requires the existence of a completely new form of matter called 'mirror matter'.

At first, it seemed too fantastic to really exist. Yet, over the last few years it appears that almost *every* astrophysical and experimental prediction of the mirror matter theory has actually been observed by observations and experiments: There is fascinating evidence for mirror matter in the Universe from astronomical observations suggesting that most of our galaxy is composed of exotic dark material called 'dark matter'. Recent particle physics experiments have revealed unexpected properties of ghostly particles called 'neutrinos' and weird matter anti-matter atoms. This unexpected behaviour is expected if mirror matter exists. Most remarkable of all is the evidence that our planet is frequently bombarded by mirror matter asteroid or comet sized objects, causing puzzling events such as the huge 1908 Siberian explosion which felled more than two thousand square kilometres of native forests without leaving a single meteorite fragment behind! Altogether I will discuss seven major puzzles in astrophysics and particle physics each arguing in favour of the mirror matter hypothesis. There are indeed seven wonders of the mirror world...

New data from current and future experiments will keep coming in even as this book is being printed. Unfortunately, I am not a fortune teller and do not know what these future experiments and observations will find. However, I can predict what they will find if mirror reflection symmetry and hence mirror matter exists. The case for mirror matter will therefore either strengthen or weaken as new data comes in and future experiments are done. In the meantime, I advise you to sit back, relax and let me take you on a journey exploring one of the boldest scientific ideas ever proposed.

No scientist works in isolation and I am no exception. I have had fruitful collaborations on mirror matter with a number of very creative people, including Sergei Gninenko, Sasha Ignatiev, Henry Lew, Zurab Silagadze, Ray Volkas and T. L. Yoon. I have enjoyed interesting correspondence on some aspects of this subject with Sergei Blinnikov, Zdenek Ceplecha and Andrei Ol'khovatov. In addition, I would like to acknowledge invaluable support over the years from many friends and colleagues including in particular, Pasquale Di Bari, John Eastman, Greg Filewood, Dave Howland, Girish Joshi, Matthew Tully, and Nick Whitelegg. I am also greatful to many of the above people, and also Jaci Anderson and Glen Deen for providing me with useful comments on the manuscript and Tony Nguyen for helping with the cover.

Of course, I thank my family most of all. It is to them that I dedicate this book.

Robert Foot
August 2001

for Carolyn and James

Contents

There are more things in
Heaven and Earth, Horatio, than
are dreamt of in your philosophy.

William Shakespeare – Hamlet.

PART I

Why Mirror Matter?

Chapter 1

Introduction

Shortly before his death in 1727, Isaac Newton reflected upon his life and wrote[1]:

> I don't know what I may appear to the world, but, as to myself I seem to have been only like a boy playing on the sea shore, and diverting myself in now and then finding a smoother pebble or a prettier shell than ordinary, whilst the great ocean of truth lay all undiscovered before me.

More recently in Stephen Hawking's *a brief history of time*, it is written[2]:

> I still believe that there are grounds for cautious optimism that we may now be near the end of the search for the ultimate laws of nature.

The contrast between the current Lucasian Professor and the former holder of that position is striking. Hawking is not alone in his prophecy. It has been repeated with monotonous regularity since the days of Maxwell (1865). One day it may come true, but that day is a long way off. I believe that a revolution in science may be imminent. In fact, over the last decade, remarkable evidence from astronomy (studies of the very big) to studies of the elementary particles (the very small) suggest that a completely new type of matter exists – 'mirror matter'. The best ideas in science are usually very simple,

3

and fortunately mirror matter belongs to this category. I believe that the ideas and the evidence can be appreciated by anyone interested in science.

In the process of uncovering mirror matter we will encounter many recent and unexpected discoveries, including:

- Invisible stars which reveal their presence by gravitationally bending the light from more distant stars behind them. I will argue that these invisible stars are made of mirror matter which can simply explain why we don't see them.

- Planets orbiting nearby stars which are *eight times* closer to their star than the distance Mercury orbits the Sun. I will suggest that these unexpected planets are expected if they are made of mirror matter.

- Bizarre, apparently free-floating planets wandering through space. They can be more naturally interpreted as ordinary planets orbiting mirror stars, but I could be wrong!

- Strange and unexpected properties of elementary particles such as the ghostly neutrinos. These particles are emitted from the Sun and in other processes. However, half of them are missing! The missing neutrinos may have been transformed into mirror neutrinos as I will explain.

- I will also discuss a strange class of 'meteorite events' such as the huge Siberian 1908 explosion and other similar such explosions. There is evidence that these explosions are caused by the random collisions of our planet with orbiting 'mirror matter space-bodies'. Most remarkable of all is the real possibility that mirror matter remnants may still be in the ground today! Needless to say the possible uses of this new type of matter are not even imagined...

By the way, this is a (generally) serious scientific book. However, unlike other 'serious scientific books' this book does not claim to reveal the 'mind of God'. In fact, not many ridiculously grandiose statements will be made at all. Rather, it is simply a book about mirror reflection symmetry – and its far reaching implications.

Symmetry is a word frequently used in everyday language and we are all aware of what it means. Examples of symmetrical objects abound: flowers, butterflies, snowflakes, soccer balls and so on... In fact, as some of these examples illustrate, symmetry is often associated with beauty and vice versa. It is perhaps not surprising then that symmetry plays a pivotal role in our understanding of the elementary particles and their forces, but let me start at the beginning.

There are many distinct types of symmetry. The symmetry of a mushroom is completely different to the symmetry of a butterfly which in turn is completely different to the symmetry of a soccer ball. A butterfly is an example of the most familiar symmetry – 'left-right' symmetry. This symmetry occurs when two equal portions of a whole are the mirror image of each other. For obvious reasons, this symmetry is also called 'mirror' symmetry. A soccer ball is an example of another type of symmetry – rotational symmetry. In fact, it is an example of an object with three dimensional rotational symmetry because rotations around any axis do not change the appearance of the ball. Finally a straight fence or railway line are examples of objects which display another type of symmetry – translational symmetry. A railway line or fence looks the same as we move along it.

Fortunately the everyday usage of the concept of symmetry is exactly the same as its technical usage in science. Although it is often useful to describe symmetry in a mathematical way – this need not concern us. Here we need only discuss the ideas and concepts which is enough to glimpse the beautiful world of the elementary particles and their interactions.

Most people are aware that ordinary matter: you, me and everything else we see, except light itself, is composed of atoms. Although atoms are very tiny, approximately one ten millionth of a millimetre in size, they are still not the most fundamental building blocks of matter. Atoms are not *elementary* entities. Each individual Atom is made up of electrons and a compact nucleus, which in turn is made from protons and neutrons. There are about 100 different types of atoms depending on the number of electrons that they contain. The science of atoms, how they interact with each other to form molecules and how different molecules interact with each other is of course the science of chemistry. However, we will not be involved so much with chemistry but with the most fundamental of the sciences – physics. One thing that physics is concerned with is the most basic questions that can be asked. For example, what are the properties of the elementary particles: protons, neutrons, electrons from which all matter is made? How do these particles interact with each other and with light?

One thing that has been learned over the years is that the interactions of elementary particles display a variety of symmetries. Some of these symmetries are quite familiar such as rotational symmetry and translational symmetry. Thus, the laws of physics remain the same whether we are in Melbourne or in Moscow, which means that Russian physics text books are useful in Australia and vice versa (after they are translated...). In addition to translations in space (and translations in language!) we can imagine translations in time. The laws of physics are the same today as yesterday or even a century ago, however our knowledge of these laws generally improves as time goes by. Hence, physics text books are not the same today as a century ago, yet the laws of physics are the same. There are still other more abstract symmetries of the elementary particle interactions. These are called 'Lorentz symmetry' and 'gauge symmetry', which are nevertheless quite elegant and natural once you get to know them.

Progress in science is rarely a smooth comfortable journey. Rapid progress generally occurs in brief intervals usually through new and unexpected experimental results and sometimes through novel theoretical ideas. Of course progress is most rapid when theory and experiment move together in harmony. One of the most remarkable theoretical ideas of the 20th century was the discovery of relativity theory in 1905 by Albert Einstein. Space and time were unified with time becoming the fourth dimension. Einstein suggested that the laws of physics were symmetrical under rotations in this four dimensional space-time, rather than just the three dimensions of space. The predictions of this theory, such as moving clocks must run more slowly, have been experimentally verified with tremendous precision. This is possible because Einstein's theory not only tells us that moving clocks run more slowly, but it tells us exactly how much more slowly! This four dimensional rotational symmetry of space-time is called 'Lorentz symmetry'[*].

There are four known fundamental forces in nature: gravity, electromagnetism, weak and strong nuclear forces. Gravity is quite familiar to most of us. It keeps our feet on the ground, it keeps our planet and all the other planets in our solar system in orbit around the Sun and it keeps the Sun in orbit around the centre of our galaxy. Electromagnetism is no less important – while it is gravity that holds us down, it is electromagnetism that stops us from falling through the floor. It is also the force responsible for electricity and magnetism. While the weak and strong nuclear forces are less familiar, they are nevertheless equally fundamental and important as the other more familiar forces. For example, the weak and strong nuclear force provides the energy which powers the Sun, without which our planet would be too cold to sustain life.

Today we know that three of these forces, electromagnetism, the weak and strong nuclear forces are, mathematically, very similar and fairly well understood. Gravity, on the other hand, is quite different and its relation to the other forces is somewhat mysterious. One reason is that gravity can be described in geometrical terms as a curvature of four dimensional space-time while the other three forces are described in terms of symmetries on an abstract 'internal' space, which is nothing to do with ordinary space-time that we know about. These peculiar symmetries of the electromagnetic, weak and strong nuclear forces are called 'gauge symmetries'.

Evidently, symmetries are rather important in understanding the elementary particles and their forces. However, it is pertinent to recall that these symmetries were not always so obvious. I have already mentioned the case of Lorentz symmetry – the rather abstract idea that space and time can be treated mathematically as a four dimensional space-time. In fact, after the discovery of relativity theory and Lorentz symmetry, an English Physicist called Paul Dirac uncovered a big problem. In the late 1920's Dirac noticed that a microscopic mathematical description of the electron consistent with

---

[*]In addition to Albert Einstein's insight, important contributions to the relativity theory were made by others, including: Hendrik Lorentz, Hermann Minkowski and Henri Poincare.

Lorentz symmetry was not possible, unless something completely new existed. Nothing short of a new form of matter was required to reconcile Einstein's relativity theory with the quantum mechanical theory of the electron. This new form of matter, called 'anti-matter' was thereby theoretically predicted to exist.

Specifically, Dirac predicted that in addition to the particles that make up ordinary matter – the electrons, protons and neutrons, anti-particles called 'positrons' (or anti-electrons), 'anti-protons' and 'anti-neutrons' should all exist. The symmetry required each type of anti-particle to have the same mass as the corresponding particle. Positrons and anti-nuclei (made from anti-protons and anti-neutrons) should form 'anti-atoms'. However, anti-particles should annihilate when they meet ordinary particles producing gamma rays (high frequency light). History tells us that experiments shortly followed which dramatically confirmed the existence of Dirac's anti-particles. First, the discovery of the positron in 1932, and later, the discovery of anti-protons in the 1950's. Anti-matter is not science fantasy but science reality. Clearly, the idea of symmetry can have remarkable implications.

This book though, is concerned not with Lorentz symmetry but with left-right or mirror reflection symmetry. Let us now briefly look at the history of this symmetry. Before 1956 physicists had assumed that the laws of physics were symmetric under left-right symmetry. This would mean that for every fundamental microscopic process that is known to occur, the mirror image process should also occur. In fact left-right symmetry is such a familiar and plausible symmetry of nature that it was never seriously questioned until various experimental puzzles began appearing in the 1950's. These puzzles led T. D. Lee and C. N. Yang to suggest that the weak nuclear force does not display left-right symmetry. They proposed an experiment to directly test the idea involving the β-decay of an unstable isotope...

At the time, most scientists didn't expect that mirror symmetry could really be broken. The prevailing scepticism was summed up by Wolfgang Pauli when he wrote in December 1956[3]:

> I am however prepared to bet that the experiment will be decided in favour of mirror invariance. For in spite of Yang and Lee, I don't believe that God is a weak left-hander.

However, Pauli was not so foolish as to let his beliefs get in the way of science. He did agree that experiments should be done to check it[4]:

> I believe in reflection invariance in contrast to Yang and Lee...
> Between believing and knowing is a difference and in the last
> end such questions must be decided experimentally.

The experiment suggested by Lee and Yang was performed in 1957 by C. S. Wu and collaborators. In this experiment a number of cobalt-60 atoms were cooled down to near absolute zero Kelvin (the lowest possible temperature) and placed in a strong magnetic field. Cobalt-60 is an unstable isotope. Ordinarily, Cobalt-60 decays emitting an electron with any direction equally likely. However, under these extreme conditions, the electrons should be equally likely to emerge from the two poles of the magnetic field – if the fundamental decay process displayed mirror symmetry. Yet, it was observed that more electrons came out from one direction than the other. If we observed only one nuclei decaying we could not say anything. Mirror symmetry does *not* mean that each single interaction or decay process is the same as its mirror image – it is not. Mirror symmetry means that the mirror image process can occur and should occur with equal probability. Therefore, by observing a large number of decays of Cobalt-60 we can easily determine whether mirror symmetry is violated. The remarkable conclusion was that the fundamental laws of physics appear to be 'left-handed'. This is really very strange. Every other plausible symmetry, such as rotational and translational symmetry, are found to be microscopic symmetries of particle interactions. Can nature really be left-handed?

Are they, the fundamental laws of physics that is, really left-handed or do they only appear to be left-handed? Remember our earlier comments about Lorentz symmetry. At one time this symmetry did not *appear* to be a symmetry at all. This was because anti-matter had yet to be discovered. Only when you have particles and anti-particles is it possible to write down a consistent microscopic theory for the interactions of the electron, proton and neutron which respects Lorentz symmetry. Remarkably, it turns out that it is still possible for particle interactions to be symmetric under mirror or left-right symmetry. Just as Lorentz symmetry required the existence of anti-matter, left-right symmetry can exist if and only if a new form of matter exists – mirror matter.

Often, it seems that nature is more subtle and beautiful then first imagined. It could be that nature's mirror is of a more abstract kind. Imagine that for each type of ordinary particle there is a separate 'mirror particle'. That is, not only do we have photons, electrons, positrons, protons etc., but also mirror photons, mirror electrons, mirror positrons, mirror protons etc. We can imagine that in nature's mirror not only space is reflected but also particles are reflected into these mirror particles. The relationship between ordinary and mirror matter is somewhat like the relationship between the letters 'b' and 'd'. The mirror image of 'b' is the letter 'd' and the mirror image of 'd' is the letter 'b'. Thus, while neither 'b' nor 'd' is symmetric (in a sense they each have the opposite handedness), together 'bd' is in fact mirror symmetric, with the two letters interchanging in the mirror image[5]. Try it with a mirror and see! Still, the mirror reflection of an object appears very similar to the original. It is perhaps not surprising, therefore, that the properties of the mirror particles turn out to be very similar to the ordinary particles. For example, the mirror particles *must* have the same mass and lifetime as each of the ordinary particles, otherwise the mirror symmetry would be broken.

In some ways mirror particles resemble anti-particles. However, there is a crucial difference. Unlike anti-particles, the mirror particles interact with ordinary particles predominately by gravity only. The three non-gravitational forces act on ordinary and mirror particles completely separately. For example, while ordinary photons (that is, ordinary light) interact with ordinary matter (which is just the microscopic picture of the electromagnetic force), they *do not* interact with mirror matter. Similarly, the 'mirror image' of this statement must also hold, that is, the mirror photon (that is, mirror light) interacts with mirror matter but does not interact with ordinary matter. The upshot is that we cannot see mirror photons because we are made of ordinary matter. The mirror photons would simply pass right through us without interacting at all!

The mirror symmetry does require though that the mirror photons interact with mirror electrons and mirror protons in exactly the same way in which ordinary photons interact with ordinary electrons and ordinary protons. A direct consequence of this is that a mirror atom made from mirror electrons and a mirror nucleus, composed of mirror protons and mirror neutrons can exist. In fact, mirror matter made from mirror atoms would also exist with exactly the same internal properties as ordinary matter, but would be completely invisible to us! If you had a rock made of mirror matter on your hand, it would simply fall through your hand and then through the Earth, and it would end up oscillating about the Earth's centre[*]. We can safely conclude that if there was a negligible amount of mirror matter in our solar system, we would hardly be aware of its existence at all. Thus, the *apparent* left-right asymmetry of the laws of nature may be due to the preponderance of ordinary matter in our solar system rather than due to a fundamental asymmetry in the laws themselves.

Do mirror particles really make the laws of physics left-right symmetric? Let us consider a simple and light-hearted 'thought experiment' involving again the Cobalt-60 decay. Imagine there was a mirror planet orbiting a mirror star in a distant part of our Universe (note that there is only one space-time – there is no 'mirror Universe'). Let's call this hypothetical planet 'Miros'. Miros is a planet made of mirror matter – atoms composed of mirror electrons and mirror protons and mirror neutrons. Miros is somewhat different to Earth though. It's a bit smaller with deeper oceans, but there is life on Miros. The people of Miros are a bit strange, they have very large feet and only have one eye – but they are very happy. They have wise leaders who would never dream of putting nuclear missiles in space and they realised very early the importance of reducing green house gases. On Miros a football team called 'Collingwood' often wins the football. Thus, Miros isn't much like Earth which just illustrates that microscopic symmetry of particle interactions does not translate into a macroscopic symmetry.

---

[*]Later I will discuss the possibility that a *new* type of interaction (or force) could exist coupling ordinary matter to mirror matter. If this is the case, it may actually be possible to pick up a mirror rock, although it would still be invisible. Clearly, the consequences of such a force are very important and it will be considered in chapter 5. However, in order to keep this introductory discussion as simple as possible, this possibility has been ignored.

Anyway, our mirror matter friends on Miros realised the importance of pure science; their wise government always made sure that financial support was given to those mirror scientists who had a research record consisting of interesting and innovative ideas. One day someone on Miros had the idea that they should test whether the fundamental laws of nature are mirror symmetric or not. So they set up their Cobalt-60 experiment with a similar experimental set up as was done by people here on Earth in 1957. But what they found was something quite different. They found the mirror image result. That is, they found that the mirror electrons were mostly emitted from the decaying Cobalt-60 mirror nucleus in the opposite direction as was found here on Earth. Our mirror friends on Miros concluded that the laws of physics were right-handed.

The laws of physics cannot both be left-handed and right-handed. Ordinary particles form a left-handed sector, mirror particles form a right-handed sector. Taken together, neither left nor right is singled out, since ordinary and mirror particles are otherwise identical. (This is much like the letters 'b' and 'd'; 'b' represents the ordinary particles and interactions and 'd' the mirror particles and interactions). However, if mirror particles don't exist anywhere in the Universe then the laws of physics are indeed left-handed. Similarly if the Universe was full of mirror particles with no ordinary ones, then the laws of physics would be right-handed, but if both ordinary and mirror particles exist together then left-right symmetry is restored.

The basic geometric point is illustrated in the following diagram.

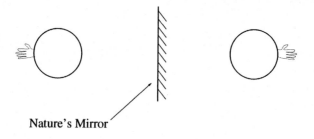

Nature's Mirror

The left-hand side of this figure represents the interactions of the known elementary particles. The forces are mirror symmetric like a perfect sphere, except for the weak interaction, which is represented

of the known elementary particles – the neutrinos, to observations of the largest systems – galaxies of stars in the Universe. After reading this book the reader will be aware of the evidence and may make his or her own judgement. At the very least, the question of the existence of mirror matter is one of the most interesting questions in science at the moment, and it should be (hopefully) answered in the next five years.

In the following chapters, I will provide the general picture of how we can find out if mirror matter actually exists and why the case for its existence currently seems so strong. There are broadly two different strategies which can be used to test the theory. First, because mirror matter is stable and behaves much like ordinary matter, it should exist in the Universe today. If one believes that the big bang theory is the correct description of the origin of the Universe, and there is some evidence for that, then mirror matter should have been created along with ordinary matter when the Universe was born. In fact, independently of whether the big bang theory is correct or not, the microscopic symmetry between ordinary and mirror matter suggests that whatever mechanism created ordinary matter should also create mirror matter. In other words, an almost inevitable consequence of the idea that the fundamental laws of physics display left-right symmetry is that mirror matter must exist in the Universe. Furthermore, like ordinary matter, mirror matter can form stars, planets and asteroid sized objects which can populate the heavens. However, such mirror stars, planets and the like would be invisible to us, since mirror matter would only radiate or reflect mirror light which doesn't interact at all with us ordinary people, and our telescopes made from ordinary matter. Thus, the first main prediction of the mirror matter theory is that invisible or dark matter should exist in the Universe.

One might think that invisible dark matter would be unobservable, and this it literally is, however there are simple ways of demonstrating in quite a compelling way that it really exists. In fact, there is a lot of astronomical evidence that the Universe is full of such invisible dark matter. In the following chapters this evidence will be presented and discussed. The evidence not only suggests that most of our galaxy is made of dark matter, but that nearby stars have mirror planets and even more remarkable, that our solar system contains mirror matter 'space bodies' (that is, asteroid sized objects) which are frequently bombarding our own planet Earth.

The other main strategy for searching for the existence of mirror matter is through the implications for microscopic processes such as particle interactions. This is because it is actually possible for *new* small forces to exist, which (like gravity) act on both ordinary and mirror matter. However, because we know that the laws of microscopic particle interactions obey certain symmetries, such as rotational, Lorentz and gauge symmetries, there are only a few possible ways in which small forces can couple ordinary to mirror matter. One possible force is a small coupling of ordinary photons to mirror photons. I will explain in subsequent chapters more precisely what this statement means, however the effect of this tiny force, it turns out, is to make orthopositronium (a weird type of 'atom' made from an electron and a positron) decay faster than we would otherwise expect – an effect which has already been observed in an experiment. A more dramatic effect is that it can make mirror matter space bodies visible as they travel through the atmosphere. They may not only be visible but may explode leading to devastating consequences. In fact, the remnants of such cosmic bodies may still be in the ground at various impact sites because the small force between ordinary and mirror matter can be large enough to oppose the force of gravity.

Another way in which microscopic forces can couple ordinary and mirror matter is through a type of 'mixing' of neutrinos. Again, I will explain more precisely what this statement means later on, however, the effect of it is to make ordinary neutrinos transform into mirror neutrinos, thereby causing them to effectively disappear. As I will discuss, there is remarkable evidence that neutrinos do indeed disappear, moreover, the rate at which they are observed to disappear is predicted precisely in the mirror matter theory.

Chapter 2

Elementary Particles and Forces

This chapter didn't appear in the first draft of this book. Including too much background material can be dangerously boring. On the one hand, I wanted to get straight into the 'interesting stuff', and on the other hand, some confusion may arise for people unfamiliar with some of the basic concepts. I have therefore included this brief summary of some of the basic 'particle physics' concepts, and also emphasised again how this is extended to include the hypothesis of mirror symmetry. Let me start by sizing up the various scientific disciplines...

Nature's distance ladder

One could define the various scientific disciplines: physics, chemistry, biology, geology, astronomy and cosmology by the characteristic distance size or *scale* involved. The distance scales cover a huge range from one ten million billionth of a centimetre – the domain of particle physics, to distances of order 100 billion light years – the size of the visible Universe. One light year is the distance that light travels in one year which is itself a very large distance – about 10,000 billion kilometres.

Mathematicians are very clever people. They quickly invented a very simple way of expressing very large numbers and very small numbers. In *scientific notation* large numbers are expressed as powers of 10. For example,

$$10^0 = 1 \ (\text{no zeros})$$
$$10^1 = 10 \ (\text{1 zero})$$
$$10^2 = 100 \ (\text{2 zeros})$$
$$10^3 = 1,000 \ (\text{3 zeros})$$
$$10^4 = 10,000 \ (\text{4 zeros})$$

In this notation the size of the visible Universe is 10^{11} light years – or 10^{29} centimetres.

We can also use scientific notation to express very small numbers, like the size of an atom – about one hundred millionth of a centimetre. One hundred millionth is the fraction $1/100,000,000$. In scientific notation, $1/100,000,000 = 1/10^8$ which is conventionally expressed as 10^{-8}. Thus, the size of an atom is simply 10^{-8} cm. With this environmentally friendly paper saving notation, we can conveniently express the characteristic distance scales of nature, see **Figure 2.1**. This figure also illustrates the concept of a logarithmic scale or simply, 'log' scale. In a log scale each factor of 10 is equally spaced, but let us not worry too much about that. Instead let's go straight to the heart of (the) matter.

What is an elementary particle?

It is essentially the same question as asking "what is everything made of?". Things around us, as well as us, are made of atoms. But what are atoms made of? At one time it was thought that atoms weren't made of anything, they were indivisible – the basic building blocks of matter. Atoms are about 10^{-8} cm in size. Eventually, it was found that there were even smaller particles. Electrons were discovered (1897-1899), which we now know are point-like down to distances less than about 10^{-16} cm. Not long after that discovery it was proposed that atoms were composed of electrons embedded in a type of jelly, or rather, plum pudding (as the English would have

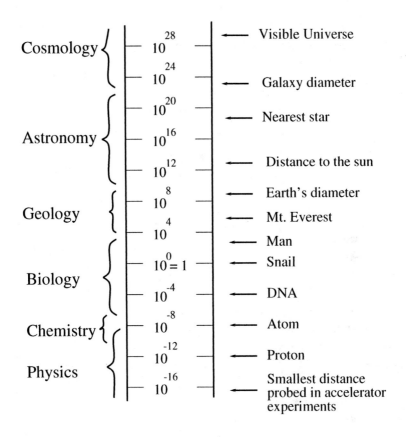

Figure 2.1: Nature's distance ladder (in centimetres).

it) of positive charge. The plum pudding model, as it was known, did not long endure. It was devoured by Rutherford in 1911.

How did Rutherford do it? Early in 1909 Rutherford suggested to two colleagues, Hans Geiger (of 'Geiger counter' fame) and Ernest Marsden to scatter a beam of α particles off a metal foil (α particles are just helium nuclei and are emitted by various radioactive elements). The α particles were very energetic, travelling at about 10,000 km/s. In the plum pudding model, these very fast α particles should just deviate only slightly when passing though the foil – but the experiment suggested otherwise. It was found that a small fraction of alpha particles actually bounced back! This means that the plum pudding was no ordinary pudding – it suggests that large pips were present. Rutherford later remarked[6]:

> It was quite the most incredible event that has ever happened
> to me in my life. It was almost as incredible as if you fired a
> 15-inch shell at a piece of tissue paper and it came back and
> hit you.

It took a short time for Rutherford to realise that the implication of these scattering experiments was that the positive charge was not spread out in a pudding, but concentrated at the centre in a 'nucleus'. Atoms are in fact mostly empty space.

Insight into the rather strange behaviour of electrons in atoms began in 1913 with the Bohr model of the atom. It was found that the traditional or 'classical' concepts were completely inadequate to describe the domain of microscopic phenomena. A completely different type of theory was needed. In short, a sort of physics revolution occurred, and by the late 1920's the quantum theory of the electron had arrived. Ambitious and occasionally even grandiose statements were heard from all quarters. One of the leading physicists of the day, Paul Dirac (the anti-matter man) proclaimed[7]:

> The underlying physical laws necessary for the mathematical
> theory of a large part of physics and the whole of chemistry
> are completely known.

The structure of the nucleus was also a great problem for many years. Most people thought that it was composed of electrons and protons – but they could never get it to work. Things were greatly clarified by the discovery of the neutron in 1932. The nucleus consisted of protons and neutrons bound together by a new force, called the strong nuclear force.

One important lesson from history is that almost all progress in science is driven by experiments. Pure thought seldom gets very far – unless it is coupled with experiments. Without experiments we might as well sit around in a hot tub and conclude that all matter consists of four elements: fire, water, air, but what's the other one? I guess it must be earth, but then what are people made of? At this point, I should probably surrender and take out my encyclopaedia, if

I had one, and try to follow the thinking of the ancients. But perhaps this procedure, apparently followed by nearly all writers of popular books on science, is missing the point. Science really began in earnest when people finally got out of their tubs and started to investigate – do careful observations, get their hands dirty and actually do experiments. Sometimes history travels in circles though. There is an interesting recent trend in particle physics. Some people are returning to their hot tubs and arguing that everything is made from 'strings' about 10^{-32} cm long which live in 10 dimensional spacetime... The idea that the world is a flat plate that sits on a tortoise is in many ways a better theory. It's a lot simpler and can be tested. Of course, we all know that it cannot explain various established things such as the fact that people who buy round-the-world airline tickets usually return safely. In contrast, string theory appears to have no testable consequences. But that's another story.

In addition to the protons and neutrons (which together are called 'nucleons' since they make up the nucleus) and the electron, one more type of particle was inferred to exist which is called the neutrino. Neutrinos are almost 'nothing'. They have no electric charge, almost no mass, and interact with the other particles extremely weakly by a new force called the 'weak nuclear force'. Yet they exist. Indeed at one time it was thought that this was all there was as far as elementary particles were concerned. In 1947 George Gamow published a book (One, Two, Three... Infinity) which summed up the situation at that time[8]:

> "But is this the end?" you may ask. "What right do we have to assume that nucleons, electrons, and neutrinos are really elementary and cannot be subdivided into still smaller constituent parts? Wasn't it assumed only half a century ago that the atoms were indivisible? Yet what a complicated picture they present today!" The answer is that, although there is, of course, no way to predict the future development of the science of matter, we have now much sounder reasons for believing that our elementary particles are actually the basic units and cannot be subdivided further. Whereas allegedly indivisible atoms were known to show a great variety of rather complicated chemical, optical, and other properties, the

properties of elementary particles of modern physics are extremely simple; in fact they can be compared in their simplicity to the properties of geometrical points. Also, instead of a rather large number of "indivisible atoms" of classical physics, we are now left with only three essentially different entities: nucleons, electrons, and neutrinos. And, in spite of the great desire and effort to reduce everything to its simplest form, one cannot possibly reduce something to nothing. Thus, it seems that we have actually hit the bottom in our search for the basic elements from which matter is formed.

Before Gamow's ink could dry, a host of new unstable particles were discovered starting with the muon in 1947, a particle which appeared to have the broad characteristics of a 'heavy electron' – about 200 times heavier in fact.

Any physical process that we observe can always be reduced to the microscopic interactions of elementary particles. In a certain literal sense, elementary particles and their interactions are at the 'heart of the matter'. In the previous chapter, I mentioned that the electromagnetic force saves us from falling through the floor (three cheers for electromagnetism!). This 'foot feat' is accomplished by the atoms in our feet repelling against the atoms in the floor – and this is not due to the odour of smelly feet! Electrons don't have feelings but do have electric charge, and anything with charge is influenced by the electromagnetic force. Like charges repel, opposite charges attract. As the electrons from the atoms in the outer surface of our feet push against the electrons from the atoms on the outer layers of the floor strong electromagnetic repulsion takes effect. This leads to a reaction force which opposes the force of gravity.

Matter Particles and Force particles

At this point it is useful to distinguish between two broad types of elementary particles. 'Matter particles' and 'force particles'. As we have seen, matter particles consist of the electron, proton, neutron (which are the constituents of atoms) and a less familiar particle called the neutrino. For each of these elementary particles there is a distinct anti-particle (which can also be classified as a type of matter particle).

Even though anti-particles are stable, you can't dig them out of the ground since they would have vanished in a 'puff of light' long ago if there was any initial anti-matter in our solar system (and maybe even in the Universe as well). Nevertheless they exist and play important roles in astrophysics, such as in supernova explosions. There are also a large number of unstable, short lived particles, which have been discovered in the late 1940's and the following decades. Each type of elementary particle has various intrinsic properties such as their mass and electric charge. They also have a certain amount of 'spin'. Roughly speaking, the elementary particles are each somewhat like a spinning top. The matter particles all have the same amount of spin, which in standard units has the value 1/2, while the 'Force particles' have twice as much spin, that is, they have spin 1.

In the 1960's - 1970's it was realised that the protons and neutrons are not really elementary. They can be viewed as being composed of more elementary constituents called 'quarks'. Quarks were first introduced by James Joyce in his book *Finnegans Wake*: 'Three quarks for Muster Màrk'. However, Joyce did not realise that the idea was more universal – not only did Muster Mark get three quarks but every proton and neutron too. This was first conjectured by George Zweig and Murray Gell-Mann in 1963. Gell-Mann's paper was rejected by the Journal *Physical Review Letters* while George Zweig's paper was only distributed as a preprint. The authorities at CERN, where he was working at the time, declared that it was too crazy to be submitted for publication.

What is it about quarks that is so crazy? The problem is that these proton and neutron constituent particles were never seen. If quarks really exist why can't we break open the neutrons and protons and isolate the three quarks? By contrast, the constituent particles of atoms are the electrons, protons and neutrons which can all be isolated. The strange behaviour of quarks was finally understood when a successful theory of the strong interactions was put together during the early 1970's. This theory suggests that quarks can never be isolated and can only exist in protons and neutrons and also in various other short-lived particles. Fortunately though, for the things that I will discuss in this book, these details are just that, details. We do not need to know anything about the detailed properties of the strong nuclear force (or indeed quarks) except for the fact that it binds protons and neutrons together into nuclei. The properties of the stable matter particles are summarized in Table 2.1[*].

Actually free neutrons are not stable, but decay with an average lifetime of about 12 minutes. However, within the nucleus they are quite stable unless the nucleus is radioactive. A radioactive nucleus is one which spontaneously decays after a certain time. There are several types of decay processes, but the one which we will be most interested in is called β-decay.

Matter Particle	Mass $\times c^2$	Electric Charge	Strong Force	Weak Force
Proton (p)	938 MeV	+1	Yes	Yes
Anti-proton ($\bar{p}$)	938 MeV	−1	Yes	Yes
Neutron (n)	940 MeV	0	Yes	Yes
Anti-neutron ($\bar{n}$)	940 MeV	0	Yes	Yes
Electron (e)	0.51 MeV	−1	No	Yes
Positron ($\bar{e}$)	0.51 MeV	+1	No	Yes
Neutrino (ν)	< 5 eV	0	No	Yes
Anti-neutrino ($\bar{\nu}$)	< 5 eV	0	No	Yes

Table 2.1: Some properties of the (stable) matter particles (and the anti-particles).

---

[*]The common unit of energy is the electron Volt, or eV. 1 eV is the energy gained by an electron after travelling through a potential of 1 Volt. 1 MeV = 10^6 eV. Also, I have expressed the mass in terms of its energy equivalent through Einstein's famous relation, $E = mc^2$. In this equation, E is the energy, m is the mass and c is the speed of light in vacuum.

In the β-decay process a neutron is converted into a proton and vice versa:

$$n \;\to\; p + e + \bar{\nu} \;\;(\beta^- \text{ decay})$$
$$p \;\to\; n + \bar{e} + \nu \;\;(\beta^+ \text{ decay}).$$

Fortunately β^+ decay is not observed to occur for free (that is, isolated) protons which can be understood from energy conservation – lighter particles cannot decay into heavier ones. Put more simply, we cannot gain weight without eating. The neutron is heavier than the proton so free protons are quite stable. However, within the nucleus things are more complicated because electromagnetic potential energy can be gained when electrically charged protons are converted into electrically neutral neutrons. For this reason it is possible for protons to decay, but only in certain nuclei. Whether or not a given nucleus undergoes β-decay, and the type of decay (β^+ or β^-) depends on the proportion of neutrons to protons within the nucleus. I will talk a little more about β-decay in a moment, but let me first introduce the notion of a force.

What is a 'force'? Macroscopically it is a type of intrinsic attraction or repulsion between objects. Without any force an object would move in uniform motion without changing its speed or its direction. Likewise when objects change their speed or direction then this is due to a force. In everyday life, we are aware of many apparently different types of forces: kicking the football imparts a force to the ball, bumping our head on the wall, accelerating in a car, etc. However, microscopically there are only four known 'fundamental' forces. They are 'fundamental' in the sense that all the other forces result from them at the microscopic level. The four fundamental forces are gravity, electromagnetism, strong and weak nuclear forces. These four forces are further distinguished by the range of their effect. Gravity and electromagnetism are long range forces – they are generally believed to exert their influence over arbitrary large distances. While the strong and weak nuclear forces are observed to be very short range – they only have a measurable effect over microscopic distances. Despite this, microscopically, the forces of electromagnetism, strong and weak nuclear forces are all

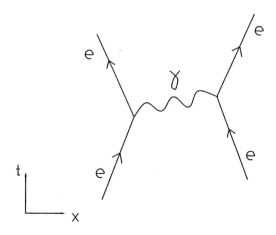

Figure 2.2: Microscopic picture of the electromagnetic force. The electron emits a photon (γ) which causes the electron to change direction. The photon is absorbed by another electron at a later time. In this and in other such diagrams, time (t) runs up the page and distance (x) runs across the page.

fairly well understood. Each force can be microscopically described in terms of the action of a 'force particle'.

Taking the electromagnetic force for example, microscopically the fundamental process involved is the interactions of electrons and protons with photons. The photon is the force particle for electromagnetism. Considering the electron, it has, at any give time a certain chance of emitting a photon. This 'interaction' causes the electron to change direction and speed if the photon is eventually absorbed by another distinct matter particle. This can be viewed diagramatically as shown in **Figure 2.2** above.

This type of diagram was first used by Richard Feynman in the 1940s. In the technical literature this type of diagram is called a 'Feynman diagram', but in this book I will use more descriptive language and call it an 'interaction diagram'. Anyway, microscopically the electromagnetic force results from photon-electron interactions.

This is the fundamental process behind the force of electromagnetism. Actually things are slightly more complicated because the exchanged photon in Figure 2.2 is not exactly the same as a real photon – it is called a 'virtual photon' in the technical literature. Again though, we don't need to bother too much about technical details such as this.

Broadly speaking, the weak and strong nuclear forces are similar to electromagnetism, but they each have certain important differences as well. Considering the weak interactions, there are not one but three force particles called W^+, W^-, Z^0. These particles were first predicted to exist in 1961 and finally discovered in an experiment in 1983. One interesting thing about $W^\pm$ particles is that when they are emitted or absorbed they always change the identity of the matter particle. For example, consider the β-decay process of the decay of a proton in the nucleus: $p \to n + \bar{e} + \nu$, which can be viewed diagramatically as shown in **Figure 2.3** below. As the diagram illustrates, the proton is converted into a neutron as it emits a W^+ particle, which later turns into a positron and a neutrino.

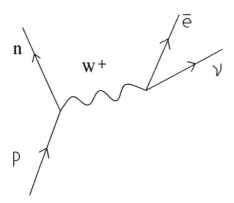

Figure 2.3: Microscopic picture of β-decay. The proton (p) is transformed into a neutron (n) by emitting a W^+. The W^+ then transforms into a positron ($\bar{e}$) and a neutrino (ν).

Turning now to the strong nuclear force, the protons and neutrons exchange not three but eight force particles called gluons. There are additional complications since these gluons are believed to interact with the point-like particles called 'quarks' within the protons and neutrons. Fortunately, complicated details such as this need not concern us at all since we only need to be aware that the strong interactions bind the quarks into nucleons and it also binds nucleons into nuclei. Nuclei of course can combine with electrons via the electromagnetic force to form atoms. Atoms can combine together to form molecules (also via the electromagnetic force) and molecules can combine together to form you and me. But of course we are perhaps more than a bunch of atoms...

The 'force particles' are summarized in the table below.

Force	Force particle
Electromagnetism	γ (photon)
Weak Nuclear Force	$W^{\pm}$, Z^0
Strong Nuclear Force	G^a (Gluons)

We don't need to know much about these details. The reader only needs to be aware that the old but good idea of forces can be viewed microscopically as due to the exchange of force particles, and that the force particle for electromagnetism is just the photon, the photon is of course the particle which makes up ordinary light.

One final comment is that gravity is not well understood microscopically. It is tempting to postulate the existence of a force particle for gravity – called the 'graviton', but the fine details are not known. What is known is that gravity is quite different from the other forces. How to reconcile gravity with microscopic physics is a deep mystery. Luckily, we don't need to worry too much about this because we only consider the effects of gravity on large objects such as asteroids, planets and stars etc. For such large objects Newton's or Einstein's 'classical' theory of gravity suffices. I will talk more about Newton and Einstein later, but for now let's return to the microscopic particle interactions of the non-gravitational forces.

Mirror Particles and Mirror Forces

The idea of mirror matter arises from the interactions of the elementary particles. These interactions are known to possess many symmetries, but the most obvious symmetry of all, left-right or mirror symmetry, is not a symmetry of the *known* elementary particles. The weak nuclear force is the culprit which is, in a sense, left-handed. As discussed in the previous chapter, this remarkable fact was first demonstrated in 1957 using β-decay experiments.

This apparent left-handedness of the fundamental laws of physics is particularly striking for neutrinos. The neutrino is an elusive elementary particle which is emitted along with the positron (or electron) in β-decay. As I have already mentioned, β-decay can be viewed as the elementary process of proton (or neutron) decay within radioactive nuclei (such as Cobalt-60), $p \rightarrow n + \bar{e} + \nu$. Like most elementary particles, such as the electron or proton, the neutrino always has a certain amount of 'spin'. I have already mentioned that this means that each electron can be viewed, roughly speaking, as a 'spinning top'. Spin is an intrinsic property like mass or charge. Every neutrino, electron or proton, always has the same amount of spin, although it may point in different directions.

A remarkable observation though is that in β-decay, or any other process that produces neutrinos, the neutrinos (ν) *always* have their spin axis orientated in the same direction relative to their direction of motion. If the neutrino were coming towards you, you'd see it spinning clockwise, in other words, it twists like a left-handed corkscrew. Just as a clockwise spinning top becomes an anti-clockwise spinning top if viewed in a mirror, the left-handed neutrino becomes a right-handed one when viewed in a mirror (see **Figure 2.4** on the following page). Thus, mirror symmetry would suggest the neutrino should be emitted with a right-handed spin half of the time. Yet, nobody has ever observed a single right-handed neutrino. In contrast, anti-neutrinos are *always* observed to be right-handed.

In the introduction I pointed out that the fundamental interactions of nature could exhibit mirror symmetry only if a set of mirror particles exist. One might wonder, though, whether mirror particles

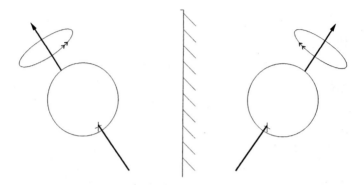

Figure 2.4: A clockwise spinning top (or left-handed neutrino) becomes an anti-clockwise spinning top (or right-handed neutrino) when viewed in a mirror.

are really necessary. Could nature's mirror reflect particles into anti-particles as well as reflecting space? This seems possible because all neutrinos are observed to be left-handed while all anti-neutrinos are observed to be right-handed. Such mirror symmetry would have many other implications. For a while this idea appeared to work. The anti-particle processes did appear to behave like the mirror image of the ordinary particle processes. However, this anti-matter mirror was shattered after just seven years. In 1964, an experiment demonstrated that this type of mirror was also broken.

The 1964 experiment involved a rather strange short lived particle known as a kaon. Because kaons live only a very short time before they decay – less than a millionth of a second – they may seem quite unimportant. Nevertheless they exist and their interactions must display the symmetries of nature, whatever they happen to be. Indeed, our current understanding of elementary particles not only describes the stable particles such as the protons and electrons, but also strange short lived particles (of which there are many) such as the kaons. Anti-kaons also exist and are distinct particles. The 1964 experiment demonstrated that kaons do not display left-right symmetry even when kaons are also reflected into anti-kaons. This amounts to nothing less than the apparent breakdown of any form of left-right symmetry in nature – or does it? The prospect that the most natural symmetry imaginable – mirror symmetry – is not a symmetry of nature, while every other obvious symmetry such as rotational symmetry and translational symmetry are indeed symmetries seems rather surprising to say the least.

Remarkably though, as I mentioned in the previous chapter and will expand upon here, it turns out that it is still possible for particle interactions to exhibit also mirror symmetry if a new form of matter, called 'mirror matter' exists. As just discussed above, having our mirror reflect particles into anti-particles as well as the mandatory space reflection simply doesn't work. It was a logical possibility but it didn't agree with experiments. If it doesn't agree with experiments it can't describe nature. Instead, imagine having a mirror that reflects every particle (including their anti-particles) into a completely new type of particle – which we might call a 'mirror particle'. In other words, I am proposing that for each type of ordinary particle, such as the photon, electron, positron, proton, anti-proton etc., there is a corresponding mirror particle *which is a distinct physical particle*.

This type of mirror seems a bit different to the one in your bathroom, since bathroom mirrors do not change the identity of the particles, or do they? Actually though, your bathroom mirror changes left-handed particles into right-handed ones, so the reflected image is, microscopically or 'quantum mechanically', composed of different particles or 'states'. It is therefore an *a priori* possibility that nature's mirror could reflect ordinary particles into distinct mirror particles. This mirror is illustrated in **Figure 2.5** (on the following page) with the mirror particles being distinguished from the ordinary ones with a prime ($'$).

The mirror symmetry interchanges the ordinary particles with the mirror particles as well as reflecting space, so that the properties of the mirror particles completely mirror those of the ordinary particles. This means that the mirror particles *must* have the same mass and lifetime as each of the ordinary particles otherwise the mirror symmetry would be broken. It also means that while the ordinary

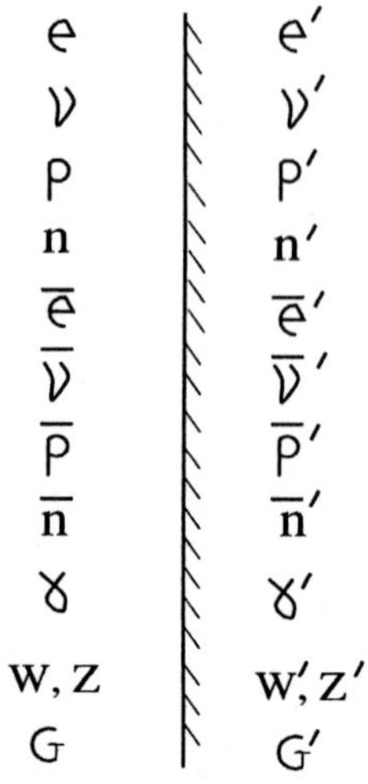

$$
\begin{array}{c|c}
e & e' \\
\nu & \nu' \\
p & p' \\
n & n' \\
\bar{e} & \bar{e}' \\
\bar{\nu} & \bar{\nu}' \\
\bar{p} & \bar{p}' \\
\bar{n} & \bar{n}' \\
\gamma & \gamma' \\
w, z & w', z' \\
G & G'
\end{array}
$$

Figure 2.5: Nature's mirror might reflect each ordinary particle into a distinct mirror particle.

particles appear in certain processes to be left-handed, the mirror particles appear in the corresponding mirror processes to be right-handed. For example, the β-decay process,

$$ p \to n + \bar{e} + \nu_L $$

where the 'L' reminds us that the neutrino is observed to be always left-handedly spinning, implies the existence of the 'mirror image' process

$$ p' \to n' + \bar{e}' + \nu'_R $$

with the mirror neutrino spinning right-handedly.

Importantly I assumed that each of the force particles also has a distinct mirror partner. This is a crucial assumption and it is necessary to explain why mirror particles are not produced in laboratory experiments. Ordinary particles interact with other ordinary particles through the exchange of ordinary force particles. Similarly, mirror particles interact with other mirror particles through the exchange of mirror force particles. There are no 'cross interactions' connecting ordinary and mirror particles from any of the known non-gravitational forces *.

Clearly, just as ordinary atoms can form by the electromagnetic force between protons and electrons, mirror atoms can form by the mirror electromagnetic force between mirror protons and mirror electrons. However, ordinary and mirror atoms do not interact with each other – except by the very feeble gravitational force. Thus, if there was a rock made of mirror matter in front of our eyes then we couldn't see it because it doesn't emit or reflect ordinary light. It could emit mirror photons if it was hot but mirror photons would pass right through us without interacting. Conversely, if we shone ordinary light on it then the ordinary photons would just pass through it. As I already mentioned in chapter 1, we couldn't pick it up because it would simply fall through our hand under the force of gravity and then through the Earth (assuming here that there are no new interactions connecting ordinary and mirror matter, see the previous footnote). We can safely conclude that if there was a negligible amount of mirror matter in our solar system we would hardly be aware of its existence at all.

Another way of illustrating the consequences of the mirror symmetry connecting ordinary and mirror particles is by considering the following 'thought experiment'. I already discussed one such thought experiment in the introduction – about a distant mirror planet called Miros. Now imagine that there is a wizard more powerful than Harry Potter, so powerful in fact that he could easily change every particle in our entire solar system into mirror particles. Would we notice? We would still be here but made of mirror atoms instead of ordinary ones, gravity would hold our feet down, mirror electromagnetism would stop us from falling through the floor (made of mirror matter) and the Sun would produce energy via the mirror nuclear force which would be converted into mirror light via mirror electromagnetic interactions. The only observable difference would be that

---

* Actually, later on in chapter 5 I will discuss the possibility of tiny *new* forces connecting the ordinary and mirror particles. However, these are new forces which are completely independent from the four known forces. For the purposes of this preliminary discussion a detailed examination of possible small forces connecting ordinary and mirror particles is ignored.

the stars in the night sky would look different – if we are made of mirror matter we would see mirror stars instead of the ordinary ones. Thus, assuming that there are both ordinary and mirror stars in the sky, we would see a different set of stars if we were made of mirror matter. The only other difference would be that in β-decay the mirror neutrinos would all be right-handed instead of left-handed...

Whose crazy idea?

Scientists often amuse themselves by arguing about the priority of ideas – everyone needs a hobby! Who did what, when. In the 17th century Newton and Leibniz had lots of fun arguing about who really discovered calculus. In the case of mirror matter the idea should date from sometime after 1956, since before this everyone generally assumed that the fundamental interactions were already mirror symmetric so there would have been no reason for postulating the existence of mirror matter. It is somewhat surprising to learn that the idea of mirror matter didn't take long to be proposed. The idea first appeared in the scientific literature in 1956, the same year that it was suggested that the ordinary interactions did not respect left-right symmetry. In fact, not only in the same year, but also by the same authors (Lee and Yang) and also in the same paper! While the Lee and Yang paper was devoted to arguing that left-right symmetry may be broken by the weak interactions of the ordinary particles, the last two paragraphs suggested that it could be unbroken if mirror matter existed. In the words of Lee and Yang (from their 1956 paper[9]):

> As is well known, parity* violation implies the existence of a right-left asymmetry. We have seen in the above some possible experimental tests of this asymmetry. These experiments test whether the present elementary particles exhibit asymmetrical behaviour with respect to the right and the left. If such asymmetry is indeed

---

*Authors Note: 'parity' is another term used in the technical literature to describe mirror symmetry. 'Parity violation' means 'violation of mirror symmetry'.

found, the question could still be raised whether there could not exist corresponding elementary particles exhibiting opposite asymmetry such that in the broader sense there will still be over-all right-left symmetry. If this is the case, it should be pointed out, there must exist two kinds of protons p_R and p_L, the right-handed one and the left-handed one. Furthermore, at the present time the protons in the laboratory must be predominately of one kind in order to produce the supposedly observed asymmetry,.....

In such a picture the supposedly observed right and left asymmetry is therefore ascribed not to a basic non-invariance under inversion, but to a cosmologically local preponderance of, say, p_L over p_R, a situation not unlike that of the preponderance of the positive proton over the negative. Speculations along these lines are extremely interesting, but are quite beyond the scope of this note.

Lee and Yang never returned to the mirror matter idea and were content with receiving the Nobel prize for their work suggesting that mirror symmetry was broken. In fact, the idea was largely forgotten with only a handful of papers written on the subject during the following three decades. My colleagues, Henry Lew, Ray Volkas and I, blissfully unaware of the last two paragraphs of Lee and Yang's paper, rediscovered the idea in 1991 and put it into a modern context. More recently, Zurab Silagadze also rediscovered the idea while reading the 'Encyclopaedia of Anomalous Phenomena' – I'll have to get a copy of that book! I have also been told by a kind Professor from India that the idea first appeared several thousand years ago in the ancient book the 'Upanishads'.....

In physics many seemingly simple and elegant ideas are proposed only to be eventually discarded when they are carefully checked by experiments and observations. This is something which distinguishes science from other disciplines. The fate of the mirror matter theory therefore rests with experiments and astronomical observations – it cannot be decided by pure thought. It is time now to examine the evidence....

*It's good to have an
open mind, but not so open
that your brains fall out.*

Bertrand Russell

PART II

Evidence for Mirror Matter in the Universe

Chapter 3

Discovery of Mirror Stars?

If mirror matter really does exist then it is reasonable to suppose that it exists in our galaxy and in other galaxies. Yet, because it is invisible, neither emitting nor reflecting ordinary light, it would be completely dark. This does not mean, though, that it cannot have observable consequences because even invisible dark matter can make its presence known to us by its gravitational effects. A famous historical example of the power of gravity is the discovery of our 8th planet, Neptune.

The discovery of Neptune

The first six planets have been observed since ancient times. The 7th planet, Uranus, was discovered by the English astronomer William Herschel in 1781 using a homemade reflecting telescope. Prior to the discovery of Uranus, the most distant known planet was Saturn, which orbits the Sun at a distance of about 1.4 billion kilometres – nearly 10 times the distance at which the Earth orbits the Sun. Uranus, it turns out, orbits at a distance of about 2.9 billion kilometers, taking approximately 84 years to complete an orbit around the sun.

The 8th planet Neptune was discovered 65 years later, but unlike Uranus, whose discovery was accidental, the discovery of Neptune was no accident. Indeed, Neptune's discovery is a rather impressive example of the power of the scientific method. This discovery is illustrated in **Figure 3.1**.

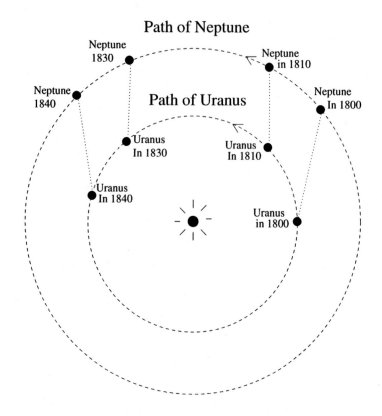

Figure 3.1: The relative positions of Uranus and Neptune during the period 1800-1840. Between 1800-1810 Uranus was moving towards Neptune and the gravitational influence of Neptune caused Uranus to travel faster. While between the years 1830-1840 Uranus was moving away from Neptune and the gravitational influence of Neptune caused Uranus to travel slower.

Newton showed us how to calculate the orbits of the planets. Put another way, they must obey Newton's laws of motion. However, Herschel's discovery of the 7th planet Uranus eventually led to something odd. Its orbit did not follow exactly the expectations from Newton. This means that either a) Newton's laws were somehow wrong[*], b) there were mistakes in the observations of Uranus,

---

[*]Actually, Einstein showed much later that they do in fact break down under certain conditions, but this was not the reason for the anomalies in Uranus's orbit.

or c) there was something new. In October 1845 the Englishman John Adams and independently on the other side of the channel Urbain Leverrier (in June 1846) proposed that an hitherto unseen planet (Neptune) must exist further from the Sun. Not only did they predict that it must exist, but the mathematics through which physical laws are described allowed them to predict its position very accurately. In fact, the two independent calculations of Adams and Leverrier agreed with each other to within 1 degree for their positioning of Neptune.

John Adams was spectacularly unsuccessful at convincing the astronomers to search for Neptune – they either didn't understand his calculations or didn't bother to. Leverrier's efforts met with more success. The night after receiving a letter from Leverrier suggesting that he should look for the new planet, Johann Galle of the Berlin Observatory found Neptune in September 1846. But Galle's job was made easy for him – he was told where to look. Clearly, Neptune made its presence known first by its gravitational effects and was later observed directly. Galle's boss, Johann Encke, who initially thought that the search was a wild goose chase (or in the case – 'planet chase') wrote to Leverrier[10]:

> Allow me, Sir, to congratulate you most sincerely on the brilliant discovery with which you have enriched astronomy. Your name will be forever linked with the most outstanding conceivable proof of the validity of universal gravitation...

Of course, in books such as this, authors such as myself are always wheeling out successful historical examples. The reader should be aware that for every theoretical success there are also many failures. The failures, however, are not usually emphasised and are often quickly forgotten*. Still, the successful cases do show

---

*One such 'failure' which has not quite been forgotten is the story of the planet 'Vulcan'. Buoyed by his 'discovery' of Neptune, Leverrier went on to argue that a new planet – Vulcan – was required to explain an anomaly in the orbit of Mercury. Searches for Vulcan failed to find it, or rather, many searches found it, but it was never confirmed. The anomaly was later explained by Einstein in 1915; Mercury's orbital misbehaviour was not due to a new planet or due to mistaken observations, but due to the modification of gravity predicted by Einstein's general relativity theory.

that at least sometimes people get things right. With this cautionary note in mind, let me now continue the story.

Distribution of matter in the heavens

In the Universe matter is not uniformly distributed. From observations we know that matter bunches together to form stars, stars bunch together to form galaxies, and galaxies bunch together to form galaxy clusters. This is our current picture of the observable Universe. Exactly how the Universe came to be like this is certainly an interesting but very difficult problem. It is a problem which is at the forefront of modern research. Needless to say it is even now not understood and that's *why* it's at the forefront of modern research. Fortunately, the most compelling arguments for the existence of mirror matter in our galaxy are essentially observation based. They do not require knowledge of the physics of galaxy formation or complete understanding of the evolution of the Universe from its beginning, assuming it has one, to the present time. Of course, such knowledge would be very useful, but we can learn much without it.

Newton's laws of gravitation are simple and powerful. Anything with mass will influence the motion of any other body with mass. The influence is greater the closer the two bodies are. Equally important is that the effects of gravity are greatest for bodies of larger mass. Of course this effect is well known to Moon walkers. Neil Armstrong could jump higher on the Moon than on the Earth. This was not just because of his great joy at being the first person on the Moon, or because he had just bought a Toyota. Rather, it was simply because the force of gravity on the relatively light Moon is much less than the relatively heavy Earth.

There is good reasons to believe that gravity is universal. The orbits of the Moon and man-made satellites around the Earth, the orbits of the planets, comets and asteroids around the Sun all obey the same universal law. Indeed, the power of Newton's laws has been quite spectacularly demonstrated with our discussion of Neptune. What about on very small distances? Small distance scales can be studied in careful laboratory experiments using a type of pendulum

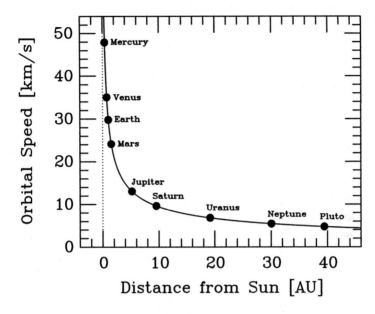

Figure 3.2: Orbital speed of the planets versus their distance from the Sun.

called a 'torsion pendulum'. Such experiments have confirmed that gravity is described by the same rules on very short distances as it is over large distances. For example, a recent laboratory experiment[11] has measured the gravitational attraction between objects just 0.2 millimetres apart, again showing that Newton's laws are upheld.

The gravitational force on the planets is due mainly to the mass of the Sun. This is because the Sun contains more than 99% of the mass of the solar system. The closer a planet is to the Sun, the greater the gravitational attractive force which the planet feels. The larger the force, the larger is the planet's orbital velocity. Conversely, the more distant the orbit the weaker the hold of gravity, which means that distant planets must move more slowly otherwise they would be flung into space never to return. **Figure 3.2** (above) shows how the orbital speed of the planets varies from their distance to the Sun. [The unit of distance is the Astronomical Unit or AU.

1 AU is the Earth-Sun distance]. As the figure shows, the velocity ranges from about 50 km/s for our closest planet – Mercury, to about 5 km/s for our most distant known planet – Pluto.

Evidently, there is a strong connection between orbital velocity and the force of gravity. In fact, knowing Newton's laws we could put constraints on the distribution of mass in our solar system by studying the orbits of the planets and the other orbiting bodies such as comets and asteroids; the mass distribution directly affects the gravitational force which dictates the orbital motion of the planets. For our solar system there is not much room for a large proportion of invisible mirror matter, or any other type of invisible matter. Any nearby mirror matter planet in our solar system would have made its presence known via its gravitational effects on the motion of the other planets or comets, in much the same way that Neptune's existence was revealed from its gravitational effect on the motion of Uranus.

Nevertheless, small bodies (for example, asteroid or comet sized objects) made of mirror matter are possible because the gravitational influence of these bodies would be far too small to have been detected. Also, a planetary or even star sized mirror object is also possible if its orbit is distant enough. In fact, in chapter 6 I will discuss fascinating evidence that there are indeed mirror matter objects out there in our solar system. There is explosive evidence that small asteroid or comet sized mirror matter objects exist and occasionally collide with the Earth as well as independent evidence for planet or star sized mirror matter objects in distant orbits from the Sun.

In any case, within the orbit of Pluto there is not much room for a large amount of mirror matter. Let us move on to larger distances. If we were to look at our solar system from a great distance away – so great that our solar system appeared as a tiny point source of light, then we would notice that it is also in motion. It is orbiting around the centre of our galaxy in a roughly circular orbit. The huge distance involved means that it orbits the centre of the galaxy only about once every 200 million years. If we move even further away, such that our galaxy was only the size of a bright point, then our solar system would be in motion around the neighbouring galaxies.

But before we go away any further, let us come back a step. While I have argued that there is no evidence for a large amount of mirror matter in our solar system (although later I will argue that there is interesting evidence for a small amount), what about on larger distance scales? Could there be a large amount of mirror matter in our galaxy? One might think that because Newton tells us there cannot be much mirror matter in our solar system, it follows that there cannot be much mirror matter in our galaxy. Still, we must be careful, the galaxy is so much larger than the solar system. It might be possible for ordinary and mirror matter to be distributed quite independently; a sort of cosmic segregation, a bit of ordinary matter here, a bit of mirror matter there...

Actually, it turns out to be very easy to understand why ordinary and mirror matter should be separated on relatively small distance scales like our solar system. However, I will postpone a discussion of this for later. In the meantime one could just keep in mind the possibility that the distribution of mirror matter and ordinary matter can depend very much on the distance scale involved. In fact, as just about any ancient person from Mongolia or Tibet would surely have testified, because their nearby region contains only land and they and nobody else they knew ever saw any oceans, the whole world must be made of land... They might have been surprised to discover that the Earth's surface is covered by more than 70% ocean...

Galaxies contain an invisible spherical halo of dark matter

It is now time to take a closer look at our own galaxy – The Milky Way. Our galaxy appears to be a typical spiral galaxy containing of order 100 billion stars which are distributed in a flat disk, with a small spherical bulge at the centre. Obviously (with current technology) we can't view our entire galaxy from a distance, but we can take pictures of other similar galaxies. **Figure 3.3** (on the following page) shows pictures of three typical spiral galaxies with different orientations.

Just like the mass distribution in our solar system could be determined by looking at the motions of the planets around the Sun, we can determine the mass of the galaxy, as well as obtaining information about the distribution of mass within the galaxy, by measuring the velocity of stars at various locations and distances from the galactic centre. We might expect that most of the mass is near the central region of the galaxy because that's where most of the light is. If this were the case then our galaxy would be dynamically similar to our solar system – but on a much larger scale. This means that the orbits of stars should show a significant *decrease* in their orbital velocity as one observes stars orbiting further and further from the galactic centre. Surprisingly though this is not the case. In fact, the velocity is more or less constant as one looks at objects with larger and larger orbits (**Figure 3.4**[12]). This is true of stars at the edge of the disk as well as stars and compact groups of stars called 'globular clusters' distributed out of the plane of the disk.

The conclusion is that there is much more to our galaxy than meets the eye. And this is in fact literally true. The mass and distribution of light emitting/reflecting matter is completely different to the mass and distribution of matter inferred dynamically from the effects of gravity through Newton's laws. The upshot is that our

---

Figure 3.3 (Top): M51: The Whirlpool galaxy is a good example of spiral galaxy with a face-on orientation. The object in the left-hand side is a separate galaxy behind the main one. (Credit: W. Keel, 1.1-metre Hall Telescope, Lowell Observatory).

Figure 3.3 (Middle): Edge-on Spiral galaxy NGC 4565. (Credit & Copyright: Richard Robinson).

Figure 3.3 (Bottom): Andromeda galaxy, which is the nearest major galaxy to our own Milky Way. It is a typical spiral with an intermediate orientation. (Credit & Copyright: Jason Ware).

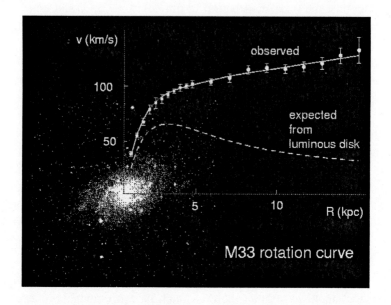

Figure 3.4: Observed orbiting velocities of stars (vertical axis) in the spiral galaxy M33, superimposed on its optical image. The horizontal axis is the distance from the centre of the galaxy in kiloparsecs (1 kiloparsecs is 3.3 thousand light years). The poor agreement between the expected velocities and the actual ones provides strong evidence for invisible 'dark matter'.

galaxy (and similar results have also been found for other galaxies) extends well beyond the visible edge. Even more interesting is that the mass is distributed spherically, roughly like a (3-D) sphere or ball, which is called the halo, despite the fact that the visible mass is predominately distributed in a flat disk.

Although this invisible mass distribution is called a 'halo', it is not much like the halo around the heads of the saints. Rather, it is a three dimensional spherical distribution which starts from the galactic centre and extends beyond the visible edge of the galaxy. The amount of mass in this three dimensional halo is estimated to be at least several times the amount of mass in the disk. Thus, in the

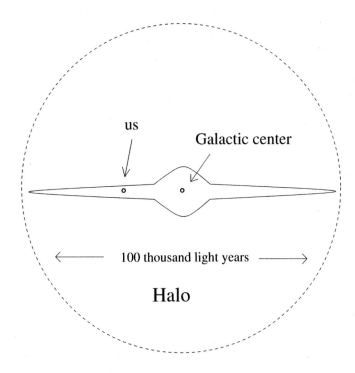

Figure 3.5: Inferred distribution of mass in our galaxy. The visible matter forms a disk, viewed edge-on in this figure, which is surrounded by an invisible three dimensional spherical halo of matter.

case of galaxies, what we see is *not* what we get. The inferred mass distribution of our galaxy is illustrated in **Figure 3.5** (above).

This result is not something that was found yesterday or even the day before. The evidence has built up over many decades; it is not even conceivable that the results could be due to mistaken observations. The observations have been repeated by many independent groups in many countries for many galaxies, all reaching the same embarrassing conclusions. We may therefore say with some certainty that either invisible dark matter exists or Newton's laws are wrong.

We know that Newton's laws cannot be completely wrong. They

have been verified for objects in our solar system with tremendous accuracy. The only known examples (in our solar system) where they breakdown are a tiny anomaly in the motion of Mercury and also the bending of light around the Sun. However, both of these examples are completely understood and were explained, or in the case of the bending of light around the Sun, predicted by Einstein in 1915. Einstein's theory of gravity goes by the title of 'general relatively'. General relativity is *not* simply a modification of Newton's laws, but a very different sort of theory. However, Einstein's general relativity theory does agree with Newton's theory when the gravitational force is not too strong, as is almost always the case in most practical examples. For this reason, Newton's laws are considered today as a very useful *approximation*. In the case of stars orbiting around the galactic centre, Einstein's theory gives the same results as Newton.

Of course, it is possible to imagine that both Newton and Einstein are wrong. Nobody is perfect. Maybe these theories only work over relatively small distances such as the size of our solar system – a mere few billion kilometres, while over larger distances they become modified in such a way to explain the motions of the stars in our galaxy without any embarrassing invisible matter. Certainly this is possible, but so far nobody has managed to find a very elegant theory which does this. Obviously, the non-existence of such a theory cannot be rigorously shown either. Nevertheless, at the present time, the most reasonable interpretation of the observations seems to be that invisible dark matter really does exist, and in fact dominates the mass of our galaxy.

The nature of the beast

If we accept that invisible dark matter really does exist, then the next logical question to ask is what is the nature of this dark matter? Is it something standard made of ordinary matter? Maybe it is in the form of faint dead stars called 'white dwarfs' or small stars that never get hot enough to burn hydrogen called 'brown dwarfs' ? Or is it something else, something more exotic?

While some of it is surely in the form of dust and gas this cannot explain the inferred dark matter in the halo. Astronomers have been able to gather information on the overall distribution of dust and gas within our galaxy by measuring the effects of the absorption of star light and by tell-tale radio (wavelength) emissions. They conclude that dust and gas contribute a negligible amount to the galactic halo, although there may be a significant component in the disk. In fact, every conventional possibility for the dark matter runs into serious problems for one reason or another.

As another example, let me briefly mention white dwarfs. White dwarfs are faint dead stars which have used up all their nuclear fuel and no longer sustain nuclear reactions. Our sun is destined to become a white dwarf one day. Currently, it is a middle aged star with no mid-life crises in sight so we don't have to worry too much at the moment. Anyway, when a star stops burning nuclear fuel its central pressure is no longer great enough to support its huge weight. The effect of this is that the star becomes gravitationally unstable. The inner part of the star collapses under its own weight with a white dwarf as the end product. Typically a white dwarf has a size as small as the Earth but with a mass comparable to that of the Sun.

Initially white dwarfs are quite hot, but since they are no longer burning nuclear fuel and producing energy, they slowly cool. However, because they are so small they are very faint. Indeed, their luminosity is proportional to their surface area which is about 10,000 times smaller than an ordinary star like our Sun. Because of their extreme faintness they can only be observed in the very nearby region of our galaxy, typically less than a few hundred light years from us (although, the youngest and hottest white dwarfs can be seen significantly further away than this). The population of white dwarfs could therefore be very numerous – perhaps numerous enough to account for the mysterious invisible mass in our galaxy. Still, there are very big problems with this idea despite its obvious merits.

The main problem with 'white dwarf dark matter' is that in the collapse process where they are formed, the outer layers of the star are ejected into space. This would lead to observable consequences which are not seen. For instance, I already discussed the fact that observations appear to exclude any significant amount of gas or dust in the halo of our galaxy. Even if the ejected gas collapses onto the galactic disk due to collisional processes, its estimated abundance would be greater than the entire inferred mass of the disk. Furthermore, this ejected material is rich in heavy elements such as carbon and nitrogen (in astrophysics any element heavier than helium is called a 'heavy element') which do not seem to be particularly abundant in our galaxy.

Other possibilities for the halo dark matter, such as black holes and neutron stars, suffer similar problems since their formation also leads to heavy element pollution and other tell-tale signs. In fact, *every conventional candidate for the dark matter is in serious conflict with observations.* Not surprisingly then, the mysterious nature of the dark matter is widely considered as the greatest of all puzzles in astrophysics at the moment.

At the end of the day we are left with the remarkable conclusion that, not only is most of the mass in galaxies invisible, but galaxies it seems are not predominately made from ordinary matter at all. Galaxies seem to be predominately made from something completely unknown, something that is, in a very literal sense, not of this world...

Enter mirror matter

Imagine that a significant part of our galaxy was indeed made of mirror matter. Could that explain the mystery of the inferred dark matter? Clearly mirror matter is dynamically very similar to ordinary matter, it would form stars, planets etc., but would not emit any ordinary light. It would emit mirror light (that is, mirror photons), but we can't detect that. In short it would be invisible – which is just what's required. So far so good. But what about the distribution? Because ordinary and mirror matter only interacts with each other via gravity, their distribution can be completely different (depending on their initial conditions such as chemical composition and angular momentum). But could this really explain why mirror matter doesn't form in a disk like ordinary matter?

Figure 3.6: M87 – an example of an Elliptical Galaxy. (Credit: Anglo-Australian Telescope, David Malin).

Perhaps a relevant piece of information is the observational fact that in some galaxies ordinary matter is distributed roughly spherically rather than in a disk. Such galaxies are called elliptical galaxies and one such example is shown in **Figure 3.6**. This immediately suggests that mirror matter could, in principle, also form a spherical distribution. So maybe our galaxy and other similar galaxies have their ordinary matter embedded into an approximately spherical mirror galaxy. But observations tell us that every spiral galaxy similar to our galaxy always seems to have an invisible approximately spherical halo. Why should it *always* occur? The fact that it always seems to occur suggests that it probably cannot be explained just from some *random* initial conditions, such as angular momentum. I feel that the answer might lie in the initial chemical make up of the Universe, as I will explain in a moment.

Galaxies are believed to be formed from a giant collection of particles held together by gravity. In other words, they were once a huge gas cloud. Within these huge systems, particles are continually colliding off each other. These collisions do two things. They create a pressure which can resist the force of gravity. If this is all that they did, the gas would never collapse; it would just sit there. However, the pressure can be reduced over time if the collisions are able to excite the atoms/molecules into higher energy levels. The excited atoms subsequently radiate photons (which eventually escape from the gas) and the atoms move back to their lowest energy state. In this way heat can be removed from the gas allowing it to become more tightly compressed, that is, to collapse.

Importantly, the collapse process occurs quite independently for the ordinary matter and mirror matter components. Why? Because collisions are an electromagnetic process, which acts independently on ordinary and mirror matter: Ordinary particles can collide with ordinary particles, mirror particles can collide with mirror particles, but ordinary and mirror particles cannot collide with each other. This means that the temperature and pressure profiles of the ordinary and mirror matter components are, in general, completely different and evolve differently. The dynamics of such a self gravitating two-component collapsing huge gas cloud is very complicated. Complicated enough perhaps to explain the vast array of galaxies and structures that are seen in the Universe.

One thing that is known though is that the way in which such a thing evolves depends quite sensitively on its initial chemical composition. Chemical composition refers to the proportions of the different elements and molecules that are present. The initial chemical composition of the ordinary matter could be quite different from the initial chemical composition of the mirror matter. Why? The answer may end up being due to the initial conditions at the very instant when the Universe was created – during the 'big bang' – and I will say a few more words about this later on. This means that the evolution of the ordinary and mirror matter components could be quite different. One could easily imagine that the rate of collapse of mirror matter into compact systems such as mirror stars/planets is much faster than ordinary matter which reduces the 'friction' between the mirror matter components in the galaxy thereby preventing collapse of the mirror matter into a galactic disk. In other words, mirror stars and planets might condense out of the primordial galactic soup before the mirror matter has time to collapse onto a disk.

It therefore seems possible for a roughly spherical halo predominately made of mirror matter to exist. It would contain mirror stars, dust and maybe also large gas clouds... But how can we test this idea?

Waiting for an exploding mirror star...

There are several ways of testing this idea that our galaxy is full of mirror matter objects such as mirror stars. First, old massive stars do not just fade away; they collapse with a bang in a titanic explosion called a supernova. These explosions are so powerful that they may even outshine the galaxy in which they appear. Such events though are quite rare, occurring in our galaxy about once every few hundred years or so. One of the most spectacular took place on the 4th of July in 1054.

Chinese observations at the time recorded that the star was so bright that it was even visible during the day-time, and was nearly as bright as the Moon at night-time. Curiously, strange lights in the sky are also reported every year in the United States also on the 4th of July, but the origin of these more recent events is undoubtly of terrestrial origin.... The remnants of the 1054 supernova explosion, known as the Crab nebula, still exists (**Figure 3.7**) and is one of my favourite astronomical pictures. [My most favourite astronomical picture is, of course, Vincent Van Gogh's *Starry night*].

The last recorded supernova event in our galaxy occurred in 1604. So we should be overdue for another... In fact in 1987 a supernova in a nearby galaxy exploded and was visible with the naked eye as well as in underground experiments, as I will explain in a moment.

How does a star get into trouble? Stars evolve peacefully for millions of years, however nothing lasts forever (including diamonds!) and eventually the star runs out of nuclear fuel. When this happens the core of the star collapses under its own weight in less than a second. If the star's mass is less than about eight solar masses then the end product is a white dwarf – an object about the size of the Earth with a mass of about the Sun. However, if the star's

The Crab Nebula in Taurus   (VLT KUEYEN + FORS2)

ESO PR Photo 40/99 (17 November 1999)   © European Southern Observatory

Figure 3.7: The Crab Nebula, which is the remnants of the 1054 supernova explosion. (Credit: FORS Team, 8.2-metre VLT ESO).

mass exceeds about 8 solar masses then the end product of this gigantic explosion is something even weirder. When such a heavy star collapses, the pressure becomes so great that all of the electrons are pressed into the protons which combine to form neutrons and neutrinos,

$$p + e \to n + \nu.$$

This is a weak interaction process, which only becomes energetically possible because of the huge gravitational pressure. Anyway, the neutrinos escape leaving behind a tightly compressed ball of neutrons – called a 'neutron star'. An object heavier than our sun, but with a radius of only about 15 kilometres. One spoonful of this material would weigh more than New York City... Clearly such material should be handled with care. If swallowed seek medical advice...

In the case of the 1987 supernova explosion, not only was the large increase in brightness observed, but the burst of neutrinos was detected in underground laboratories in Japan and the USA. A total of 20 neutrino events over a time scale of just 12 seconds were recorded. As I will discuss in later chapters, neutrinos are expected to be a window into the mirror world. A tiny mixing force is allowed and would have the effect of changing half of the supernova neutrinos into mirror neutrinos. Unfortunately, because of various large uncertainties, the initial number of supernova neutrinos that would be expected to arrive at the Earth cannot be precisely determined, so we can't tell whether half of the neutrinos are missing. This unsatisfactory situation may improve in the future with the great advances in neutrino detectors that now exist.

Even more interesting, though, is the implications of the conversion of mirror neutrinos into ordinary neutrinos which would happen if a *mirror star* explodes. This could make a mirror supernova effectively 'visible' even if no visible light is emitted. It would be a phantom explosion detectable only with neutrinos [*]. (Actually another possibility, which I will discuss in chapter 5, is that a tiny force allowing for photon mirror photon transitions could make mirror stellar explosions produce an observable burst of photons as well as a neutrino burst).

Although there is as yet no evidence of such phantom stellar explosions this is not unexpected since we know that nearby ordinary supernova explosions are relatively rare events. The estimated rate

---

[*]It is possible that an ordinary supernova explosion could mimic an exploding mirror star if the light from the ordinary supernova were blocked out by interstellar dust. However, even this possibility can be tested by looking at the direction of the neutrinos. The direction of the neutrinos from an exploding ordinary supernova should come from within the disk of our galaxy (if it happens to be one of the few halo ordinary stars it certainly won't be obscured by dust). On the other hand, the neutrinos from an exploding mirror star will most likely come from the halo, that is, in a direction out of the plane of the galactic disk.

of ordinary supernova explosions is roughly once every few hundred years – so it may not be surprising if the rate of mirror supernova explosions also turns out to be low.

Discovery of mirror stars?

On quite a different tack, even invisible stars can reveal their presence through their gravitational effects on light. In 1986, Bohdan Paczynski had a good idea[13]. His idea was to mount a search for dark matter based on the idea that a massive object could act as a sort of lens. Even if we can't see the massive object, its gravity can bend the light coming from a more distant star around it, in much the same way that light gets bent as it passes through a magnifying glass. If there are invisible bodies floating in the halo of our galaxy it is possible that they should pass between us and our line of sight to a background star. If this happens then the gravity magnifies the light from the background star as the light passes around the invisible object – causing the background star to brighten. This effect is illustrated in **Figure 3.8**.

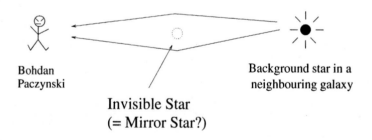

Bohdan
Paczynski

Background star in a
neighbouring galaxy

Invisible Star
(= Mirror Star?)

Figure 3.8: Magnification of a star's light by the gravity of an invisible 'star' (Mirror star?). The light bends around the invisible star just like light bends in a magnifying glass.

Because the invisible star, the background star, and our solar system, are all in relative motion, the magnification can only last for a finite time. The more massive the invisible object (the lens), the stronger its gravitational effect and the longer the period of increased brightness. For masses in the range from planet size to several times the Sun's mass, the brightness lasts from a few hours to as much as a year or so. However, even if the halo of our galaxy was full of invisible massive objects, the chance that one of these objects would pass between us and a particular background star is very low. In fact, it can be estimated that the chance is about one in a million, but the odds can be improved by simultaneously monitoring a large number of stars over several years.

Four teams of researches began the search nearly a decade ago. First off the blocks was the French collaboration Experience de Recherche d'Objects Sombres (EROS), which was closely followed by the Optical Gravitational Lensing Experiment (OGLE), run by Paczynski and colleagues at the University of Warsaw in Poland, the large Australian - US Massive Compact Halo Objects (MACHO) project, and a smaller French effort called Disk Unseen objects (DUO). All four use ground based telescopes and a large amount of computer memory to store the brightness measurements of millions of stars. When these projects started it was expected that the most likely candidate for the invisible halo objects were low mass stars called brown dwarfs weighing less than 10% the mass of the Sun. Such lightweights would be too faint to see and thus might be the invisible component of our galaxy. However, what they found was not lightweights but stellar-weight objects, with a typical mass of about half the mass of the Sun, and enough events to account for nearly half the estimated mass of the halo. Several such events are shown on the following page in **Figure 3.9** obtained by the MACHO experiment[14].

Thus, instead of finding what they most expected, the observations were able to exclude lightweight objects from being a significant component of the halo. More interesting though, is that the results can be viewed as evidence for mirror matter, since mirror stars should have a typical mass similar to that of ordinary stars – which is close to about half the mass of the Sun. In other words,

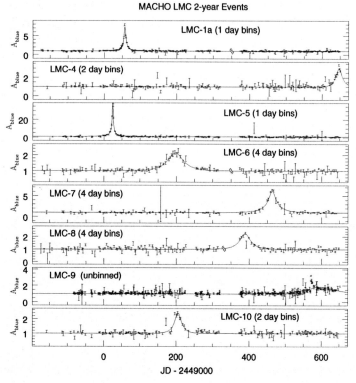

Figure 3.9: Eight distinct 'MACHO events'. The figures show the intensity of light from a background star (vertical axis) as a function of time. The light from the background star gets amplified for a certain duration, which occurs because an invisible star passes in front of the background star. In each of these cases the background star is in the Large Magellanic Cloud (LMC) – a nearby galaxy.

the data is roughly consistent with what you could expect with a mirror matter halo. The MACHOs are just mirror stars, or perhaps mirror white dwarfs... Put another way, MACHOs are really Massive Astrophysical Compact Halo *Mirror* Objects (MACHMOs).

Even more remarkable is the inferred total mass of the MACHOs found in the experiments. The outcome is that the MACHOs make up only half or a bit less of the inferred mass of the halo[15]. Recall our earlier discussion – the mass of the halo could be inferred from the motion of stars at various distances from the centre of the galaxy. The fact that the results from the MACHO experiments falls short in accounting for all of the mass of the halo suggests that there must be another component to the halo which does not show up in the MACHO experiments. Actually this feature can also be plausibly explained by the mirror matter theory. Mirror stars don't make up the entire halo simply because mirror matter, like ordinary matter, exists in two forms: In the form of mirror stars which are the MACHO events obtained in the experiments, as well as in the form of mirror dust and gas which do not leave any observable signal. [These particular experiments were only sensitive to *compact* systems such as star-sized objects, while the gravitational effect of clouds of gas and dust would be too dispersed to have been observed].

While plausibly explaining the results of these experiments is one thing – rigorous proof is another. Obviously it is difficult to rigorously prove that the MACHOs 'observed' in the experiments are mirror stars (unless one of them happens to explode in our galaxy, leading to an observable burst of neutrinos). On the other hand, if we take the mirror matter theory seriously, MACHOs are predicted to exist, and if they really are the dark matter then the results of the MACHO experiments really had to find a positive result for MACHOs about half the mass of the Sun. The fact that the results from the MACHO experiments are consistent with this prediction is strong evidence for the theory. Still, if it was the only evidence for mirror matter, then the case for its existence would be far from compelling. In short, it would be nothing to write home about – let alone a book! However, I will identify seven major puzzles in astronomy and particle physics, each plausibly suggesting that mirror matter exists.

MACHOs or WIMPs?

Let us finish this chapter with a brief discussion of the main alternative model for the dark matter. While mirror matter can lead to the formation of mirror stars, which are an example of a Massive Astrophysical Compact Halo Object (MACHO), the main alternative candidate for the invisible dark matter is appropriately called WIMPs. WIMPs are hypothetical Weakly Interacting Massive Particles. Their invisibility arises because it is assumed that they don't couple to photons. In fact, they interact only by extremely weak short-range interactions and consequently they seldom collide with each other (or with ordinary matter) which means that they can't collapse to form star sized objects. They cannot therefore explain the MACHO events. Actually they appear to have great difficulty in explaining many of the specific observations on the nature of dark matter. For example, WIMP dark matter makes specific predictions for the density profile of dark matter in galaxies which seems to be in strong disagreement with observations[16]. Furthermore, WIMPs seem unable to explain the inferred complexity of dark matter. Of course it is possible that I may be biased! In defence of WIMP theories one can say that they are very popular among particle physicists.

In view of their popularity, many experimenters have been searching for WIMPs for a long time. The idea is that these weakly interacting particles could make an observable signal in purpose-built underground detectors. Instead of going into the boring technical details, let me say something about the flavour of the WIMP search. Even better, let me quote one of the WIMP enthusiasts themselves. Twelve years ago (1989), L. Krauss eloquently captured the excitement of the hunt when he wrote in colourful language[17]:

> You are a graduate student in physics. It's late Saturday night and you would much rather be at a party. Instead, you are a mile underground, in a cavernous enclosure, entertained only by the sound of a cooling fan whirring in the desk-top minicomputer that is monitoring pulses received from the gargantuan device located in the main chamber next door. It has been a boring eight-hour shift and you long to take the elevator ride up the mine shaft to the surface, to breath the fresh air and to watch the night sky, the stars twinkling, and the cool, evanescent glow of the moon bathing the earth's surface. You are, after all, studying to be an astrophysicist, not a geologist. When you forsook a lucrative programming job in order to return to graduate school, you envisioned working at a huge radio telescope aimed at the heavens, sensing the faint pulses

emitted by quasars billions of light-years away. Yet here you are, deep underground, monitoring a new experiment built by a collaboration among four universities located on three continents. In order to pass the time you watch the calibration pulses appear with clocklike regularity on your monitor, noting how each exactly reproduces the last.

Suddenly, almost too fast to sense, you notice something momentarily different about the signal. You halt the on-line output on the computer and call up the program that single-steps through the data. While the program loads on the machine, your mind races. There is a small chance that the pulse you saw, or imagined, is the infinitesimally small signal from an elementary particle making up a totally new type of matter never before observed on earth that interacted in your detector. If so, this could be the first time this particle has interacted in the ten to fifteen billion years since it was created in the fiery Big Bang. You may be looking at a signal from the beginning of time! Such particles may constitute one hundred times more material, by weight, than everything we can see put together, thereby governing the structure, evolution, and eventual fate of the universe! Your discovery could affect the way we think about the universe as dramatically as had Copernicus's assertion that the earth moves about the sun...

Or perhaps it is just a bit of noise in the detector...

This was published in 1989, and unluckily for that poor graduate student, it was in fact a 'bit of noise in the detector'; he's still down that mine shaft waiting for a signal from the dawn of time. He's long given up hope of any excitement and regrets not taking that lucrative programming job...

Despite more than a decade of dedicated searches, no WIMPs have been detected. While I suspect that WIMPs do not exist, I definitely support such experiments – so long as I don't have to do them! It's the only way of knowing for sure. Experiments are the essence of science...

Beauty and the Beast

The most popular manifestation of WIMPs comes from a particle physics theory called 'supersymmetry'. Supersymmetry is one of those good ideas which does not seem to be used in nature. Supersymmetry is a symmetry which connects each of the known types of elementary particle with a hypothetical 'super partner' of a different spin. That such a symmetry can exist is quite non-trivial. For the mathematically minded supersymmetry holds much charm. However, its implementation as a symmetry of particle interactions is very troublesome for experiments. Most important is that the supposed 'superpartners' of each of the known elementary particles must have the same mass as the ordinary particles. This feature is very similar to the properties of mirror particles or anti-particles; they also have the same mass as their corresponding ordinary particles. That's what the symmetry tells us. The problem with supersymmetry is that if they did have the same mass, then the superparticles would have been experimentally discovered many years ago in laboratory experiments. There is simply no known way of making them invisible in the Lab – except by breaking the symmetry. It is very sad. Supersymmetry is probably what Thomas Huxley had in mind when he wrote, 'The great tragedy of science: the slaying of a beautiful hypothesis by an ugly fact', or maybe not.

Although it is possible to write down theories with hypothetical superparticles which have 'broken symmetry', they tend to be very complicated because there are essentially unlimited ways of breaking the symmetry. The resulting construction is called the 'minimal supersymmetric standard model', which has more than 100 free parameters – and that's the minimal model! Nevertheless, supersymmetry is very popular among particle physicists because they can write lots of papers predicting all of the experimental effects that these 100 parameters allow. Anyway, because of all these parameters it is possible to arrange things so that the lightest supersymmetric particle is neutral and stable, and can therefore be the dark matter of the Universe. It seems to me though, that this scenario is not very compelling because it is *ad hoc*. For example, there is no theoretical reason for any of the supersymmetric particles to be stable once the symmetry is broken. Overall, it has always seemed to me that supersymmetric models are very ugly, principally because they are so complicated and arbitrary.

This is in sharp contrast to the manifest beauty of mirror symmetry which can be completely unbroken if mirror matter exists. The microscopic properties of the mirror particles are then completely fixed without any problems for existing experiments. One could also imagine breaking mirror symmetry by giving the mirror particles heavier or lighter masses – but this leads to seven years of bad luck! I should know, I toyed with such a model in 1994, exactly seven years ago. Of course, beauty is not necessarily the same as truth. Beauty always involves some subjective judgement. In the end we must each follow our own judgement, the truth of the (mirror) matter will be decided by careful experiments and observations.

* * *

If mirror matter does indeed exist in our galaxy, then binary systems consisting of ordinary and mirror matter should also exist. Although systems containing approximately equal amounts of ordinary and mirror matter are unlikely due to, for example, the differing rates of collapse for ordinary and mirror matter (leading to a local segregation of ordinary and mirror matter), systems containing predominately ordinary matter with a small amount of mirror matter and vice versa, should exist. Interestingly, there is remarkable evidence for the existence of such systems coming from extrasolar planet astronomy, the subject of the next chapter.

Chapter 4

Discovery of Mirror Planets?

One of the most fascinating discoveries of the late 20th century was the discovery of planets around nearby stars – called 'extrasolar' planets. That nearby stars have planets is perhaps not all that surprising. What is surprising though is the bizarre and unexpected properties of these planets. As I will discuss, the properties of many of these planets suggest that they might not be ordinary planets at all, but are in fact mirror worlds.

Why search for new worlds?

There are hundreds of billions of galaxies in the observable Universe, with each galaxy containing hundreds of billions of stars. That's a lot of stars. Is our star the only one which has a planet with intelligent life? Of course nobody knows for sure. There are even some arguments about whether our own planet does indeed contain intelligent life. Such profound and weighty questions have motivated the search and discovery of planets around nearby stars. Whatever one's opinion about the likelyhood of finding ET, there is a general consensus that the discovery of ET would be the most profound scientific discovery ever made... Yet, I feel that there are more pressing reasons to search and study planets around nearby stars. Such studies will inevitably lead to a better understanding of our own solar system, the formation and evolution of other such systems, and thereby shed light on our place in the Universe. I personally am curious to know whether or not other solar systems are similar to our one. This is an interesting question because it can now begin to be answered. We have the technology!

The discovery of planets around nearby stars

The reader may well be wondering how one would go about searching for planets around other stars. Direct detection is quite difficult because a planet is a very faint object with a very small angular separation from a much brighter object – its star. So indirect techniques are used. Attention is focussed not on the planet directly but on the motion of the star caused by the gravitational influence of the planet. While the gravitational influence of a star on a planet causes the orbital motion of the planet around the star, the smaller gravitational influence of the planet on the star causes the star also to move. This effect is illustrated in **Figure 4.1**.   Clearly, if we can detect the motion of the star, which should be periodic with the same orbital period as the planet, then we can infer the existence of the planet and determine some of its properties without actually seeing it.

There are two complementary ways to detect the wobble of the star which is caused by an orbiting planet. One is by direct observation of the apparent position of the star – a technique called 'astrometry'. However, because even the closest stars are so far away, the small periodic displacement of the position of a star is very difficult to measure. For example, the size of the wobble of our sun caused by Jupiter is estimated to be 1.6 milliarcseconds when viewed from a distance of about ten light years. [1 milliarcseconds is just $1/3,600,000^{th}$ of one degree!] Ten light years is the typical distance of our nearest stellar neighbours – our nearest star is Alpha Centauri which is just over 4 light years away. Because the apparent size of the wobble obviously gets even smaller when viewed at a greater distance, the astrometric technique is best for nearby stars.

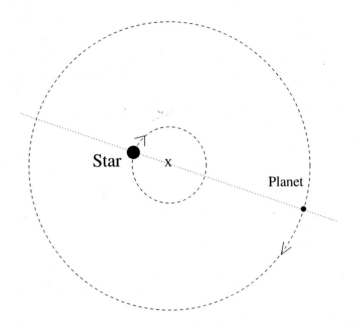

Figure 4.1: Gravity causes the planet to orbit in a circle (or more generally an ellipse) but gravity also influences the star as well. It moves in a smaller circle as shown. The centre of these circles is the point 'x', and it is called 'the centre of mass' of the system.

It is also more sensitive to heavy planets with large orbital distance because these features induce a larger stellar wobble.

The most successful technique used to date is not the astrometric one but the more subtle 'Doppler shift' technique. If the star is really moving in a periodic way then the emitted light undergoes frequency shifts. This effect, known as the 'Doppler effect', is familiar to anyone who has heard a fire engine speed by. This effect was first discussed by Christian Doppler in 1842 even before the advent of fire engines! As the fire engine moves towards us, the siren has a higher pitch (that is, higher frequency) and when it moves away from us, the sound has a lower pitch. In fact, if we imagine a fire engine moving around in a circle, perhaps the driver is confused and is reading the map book, then the frequency of the siren would be

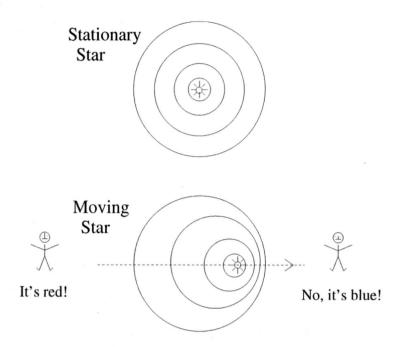

Figure 4.2: The Doppler effect. The top half of the figure shows a stationary star (with the outgoing light wave crests illustrated). The frequency
and wavelength of the light is the same in all directions. The bottom half
of the figure shows a star moving to the right. For the observer on the right
the light has shorter wavelength (which means it's shifted towards the blue)
while the observer on the left sees the light shifted to longer wavelengths
(towards the red).

highest when it is moving towards us, and lowest when it is moving
away from us, and when it is moving sideways, the frequency is the
same as if the truck was stationary. The Doppler effect is illustrated
in **Figure 4.2**.

   While light is not sound, it is well established that this effect
also occurs for light. Thus, by carefully measuring the frequency
of the light, any motion in the direction towards us or away from
us, the so-called 'radial' direction, can be measured. The Doppler
shift technique is most sensitive to heavy planets with *close-in*

orbits since these features lead to a faster moving star which produces a greater frequency shift. This is in contrast to the astrometric technique which prefers heavy planets in *distant* orbits. Another important difference is that the astrometric effect is larger for nearby stars while the Doppler shift is independent of the distance, since the latter effect depends only on the velocity of the star. Nevertheless, nearby stars are still advantageous because they are brighter, making observations easier.

During the 1990's two groups of astronomers, one in Switzerland and the other in USA began systematically searching for extrasolar planets using the Doppler shift technique. However, because the sensitivity of their measurements could only probe planets heavier than Jupiter (which were expected to be in orbits quite far from their star) the orbital period of the star's motion should be quite long. In the case of Jupiter its orbital period is nearly 12 years. Thus, to really demonstrate the existence of periodic shifts in the frequency spectrum, and thereby infer the existence of planets, great patience seemed to be the order of the day. Remarkably, in a paper[18] published in 1995, two Swiss astronomers named Michel Mayor and Didier Queloz announced the startling discovery of the first extrasolar planet, orbiting around the star 51 Pegasi. Even more remarkable was that this discovery was announced after less than a year of data taking.

Brave New Worlds

Queloz and Mayor were late starters in the field of extrasolar planet hunting. They began their planet hunt in April 1994. By that time, several other groups had been searching for new worlds, but for one reason or another, were making rather slow progress. However, Queloz and Mayor used a very simple and very clever experimental setup, allowing them to read off the Doppler shifts, with typical Swiss precision, just minutes after they observed each star.

One night in September 1994 they first set their sights on 51 Pegasi. Over the next few months the puzzling and unexpected behaviour of 51 Pegasi was revealing itself. At first, they thought that their equipment wasn't working properly. There were large changes to the Doppler shift from one day to the next. How was that possible? Their equipment was only sensitive to large Jupiter-mass planets. But these should be in a distant orbit where the Doppler shift changes very slightly from one day to the next. Eventually, after carefully checking their apparatus and the periodicity of their signal, Mayor and Queloz realised exactly what they had discovered. They had not found a Jupiter-mass planet orbiting its star with a period of several years, as most of the 'experts' had foretold, but rather, a system with a Jupiter-mass planet with orbital period of just 4 days! The Doppler shift measurements of a similar system discovered in the year 2000, which has an even shorter orbital period of about 3 days[19], are given in **Figure 4.3**. Even the closest orbiting planet in our solar system – Mercury – takes about 90 days to go around the Sun, and Mercury is a small rocky planet not a large gas giant.

Geoffrey Marcy of the USA team, who missed out on the pioneering discovery but together with Paul Butler has since discovered more extrasolar planets than anyone else, said later[20]:

> In retrospect, we feel like fools. The really laughable thing is that we believed the theoreticians. They agreed that Jupiter and Saturn-like planets form far from their host stars with long orbital periods of 10 or 20 years. So we thought we could spend 10 years working on our optics, hardware and software. Of course, 51 Pegasi proved that Jupiter-like planets could exist close to their parent star, in orbits of just a few days. With that realisation, we knew that sitting in our data, literally on hard disc, would be planets in shorter periods than the theoreticians had predicted.

And so there was! Within a few months Butler and Marcy were able to announce of few of their own...

We know a lot about nearby stars. We can measure their distance accurately and learn about the stars from the luminosity and frequency spectrum of the light they emit. The star 51 Pegasi is a

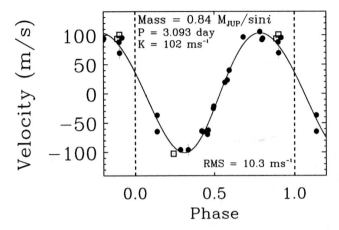

Figure 4.3: Doppler shift measurements for the close-in extrasolar planet HD179949. The peaks occur when the star moves towards us while the dip occurs when it is moving away from us. The period is just 3 days!

fairly typical solar type star, and its mass can be inferred to be approximately the same as the mass of our sun. Knowledge of the mass of the star and the orbital period of the planet is enough information to work out the distance of the planet from its star given that we know Newton's laws. It turns out to be just eight million kilometres for the planet around 51 Pegasi. This is close by any measure. For example our Earth orbits at a distance of about 150 million kilometres and even our closest planet, Mercury, keeps its distance – about 60 million kilometres. Another relevant scale is the size of the star itself, a diameter of about 1.4 million kilometres.

The amplitude or size of the frequency shifts also provides important information. The amplitude allows a measurement of the minimum mass for the orbiting planet. More precisely, it allows a measurement of the product $M_{planet} \times \sin\theta_I$, where $\sin\theta_I$ ranges from zero for a planet orbiting with a face-on orientation (that is, in the purely transverse direction, across our line of sight to the star) to the value 1 for a planet orbiting with an edge-on orientation (that is, in the purely radial direction, along our line of sight to the star).

Since the discovery of 51 Pegasi, these startling results have been confirmed by other groups, and a large number of similar systems have now been discovered. The total number of extrasolar planets now stands at about 70. Roughly half of them are of the close-in variety. All of these planets have revealed their presence because of the periodic tug of gravity on their parent stars leading to observable Doppler shifts. Table 4.1 lists the closest of the close-in planets, which I define as those with orbital radius less than 0.2 AU. [Recall that 1 AU is the distance with which the Earth orbits the Sun].

The results were so surprising that David Gray of the University of Western Ontario argued[21] that the planets, or at least the one around 51 Pegasi, may not really exist. After all, nobody has ever seen one. Perhaps the surface of the star was oscillating in and out in such a way as to mimic the effects of the orbital motion of a planet? However, such an oscillation effect would leave various other telltale signs which were searched for and never found. Still, it would be nice to detect these planets more directly or with another independent technique. The extraordinary nature of the extrasolar worlds really demands confirmation from another type of observation so that no shadow of doubt can remain. The clincher came one dark but clear night in September 1999...

A transit in the night

The reality of these close-in planets was put beyond any conceivable doubt by the discovery of a transiting system. Every time a close-in planet is discovered there is a small but significant probability that the orbit of the planet is approximately edge-on when viewed from Earth. In such an orbit, the planet will periodically pass between its star and our line of sight to its star. When it does this, it would block out some of the light from the star, thereby making the star slightly dimmer as the planet transits across the star. This is roughly similar to the way in which the Moon blocks out the light from the Sun during an eclipse, although in the Moon's case, it manages to block out all of the light.

STAR	Mass $(M_{Jupiter} = 1.0)$	Radius (AU)	Period (days)
HD83443	0.35	0.038	2.99
	0.16	0.17	29.8
HD168746	0.24	0.066	6.41
HD46375	0.25	0.041	3.02
HD108147	0.34	0.098	10.88
HD75289	0.42	0.046	3.51
51 Pegasi	0.47	0.05	4.23
BD-103166	0.48	0.046	3.49
HD6434	0.48	0.15	22.1
HD187123	0.52	0.042	3.097
HD209458	0.69	0.045	3.52
Upsilon	0.71	0.059	4.617
Andromedae	2.11	0.83	241.2
	4.61	2.50	1267
HD192263	0.76	0.15	23.87
HD38529	0.81	0.13	14.4
HD179949	0.84	0.045	3.09
55Cnc	0.84	0.11	14.6
HD13022	1.08	0.088	10.7
HD217107	1.28	0.07	7.11
Gliese 876	0.56	0.13	30.1
	1.98	0.21	61.02
HD195019	3.43	0.14	18.3
G186	4.0	0.11	15.8
Tau Bootis	3.87	0.046	3.31
HD98230	37	~ 0.06	3.98
HD283750	50	~ 0.04	1.79

Table 4.1: Listed is the minimum mass (in terms of Jupiter masses), orbital radius (more precisely semi-major axis), and period (in days) of all of the confirmed planets with orbital radius less than 0.2 AU. Also listed are stars which have multiple planets. For the latest information on extrasolar planets, see the Extrasolar Planet Encyclopaedia: http://cfa-www.harvard.edu/planets/encycl.html

For planets such as 51 Pegasi the chance of an edge-on orbit would be roughly 10%, that is, one chance in 10. While the odds are low, they are definitely large enough to make it worthwhile for astronomers to search for it. The amount of light that would be expected to be blocked out should be simply the ratio of the cross sectional area of the planet to the cross sectional area of the star, which is easily estimated to be a few percent for a planet the size of Jupiter and a star the size of our sun. In other words, if a transit did occur then the light from the star should be a few percent dimmer which is easily detectable.

Of course, even if we had a transiting system, that is, a star with an edge-on orbiting planet, we could only observe the dimming effect of the transit if we happened to be looking at the star precisely when the planet passed in front of it. Fortunately, things are made quite easy because the Doppler shift measurements tell us *exactly* when to look! A transit can only occur when the planet is directly between us and its star which means that it only has sideways (or transverse) motion and hence no Doppler shift at this point. Thus, from a Doppler shift analysis, such as the one shown in Figure 4.3, the time when a transit should occur can easily be obtained (if the planet has an edge-on orbit). Furthermore, because the orbital period is unchanging, the time when the transit occurs can be predicted well into the future, just like solar eclipses.

Clearly, knowing the time when a transit should occur makes it very easy to test for it. After every planet is discovered by the Doppler shift technique a transit search is also done. The first few systems had no observed transit. While this was not particularly surprising since each close-in system has only about a 10% chance of a transit, it did at least allow some possible doubt to remain. However, planet number 13 was the lucky one. It orbits a star which happens to have the rather inconspicuous name of HD209458.

The transit of HD209458 was observed by two groups. The first group comprising Tim Brown, David Charbonneau, David Latham and Michel Mayor observed the transit[22] during September 1999. These authors quietly submitted their paper for publication in the *Astrophysical Journal* but otherwise did not publicise their findings. Their results are shown in **Figure 4.4**. More accurate results of the

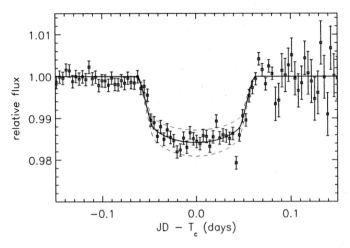

Figure 4.4: Observations of the light from the star HD209458 which has a close-in transiting planet. The dip in the intensity is due to the passage of the orbiting planet which blocks out the light as in passes in front of the star.

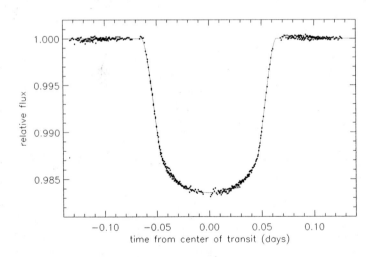

Figure 4.5: Another observation of the transit of HD209458 (the same system as Figure 4.4) but this time with much greater accuracy. These observations were taken with the Hubble space telescope.

same system were later taken with the Hubble space telescope[23] (**Figure 4.5**).

Several months later, another group led by Greg Henry[24] also observed the transit and immediately publicised their results. I remember reading about their discovery on the front page of the *Melbourne Age* (my local newspaper) in November 1999, literally only a week or two after their transit observations. Quite exciting stuff. Actually, the second groups' results were not as nice as the first group, since they only had half a transit....unfortunately some clouds got in the way at the crucial time, and they were unable to immediately obtain the full transit. Still, half a transit is better than no transit... An artists impression of the transit of HD209458 is shown in **Figure 4.6**.

Besides rigorously proving that extrasolar planets exist, the observed transit allowed for an accurate determination of the size and mass for this system. The size can simply be obtained by measuring exactly how much light gets blocked out, while the fact that the planet transits its star means that we know the alignment of the orbital plane of the system – it is edge-on ($\sin \theta_I \simeq 1$). This last piece of information, coupled with the Doppler measurements is enough to determine the mass of the planet. In the case of the transiting system HD209458 its mass is measured to be $0.69 M_J$ with a radius of about $1.3 R_J$ (where M_J, R_J are the mass and radius of Jupiter).

The reader may have guessed my own interest in the extrasolar planets. To understand what they may have to do with mirror matter I need to back-track a little and explain why exactly it was that people thought that large mass planets couldn't exist so close to their star.

How solar systems and planets are formed

Star formation is relatively brief, taking only a few million years. Unfortunately, we cannot hope to observe the whole process from start to finish for a single system – even with great advances in medical science! However, by observing many different systems at different stages in their evolution we can obtain a picture describing how star formation proceeds.

Figure 4.6: Artist's depiction of the transit of HD209458. (Credit & Copyright: Lynette Cook).

Stars form from the interstellar medium – the thin gas with a trace of dust that pervades interstellar space. It is predominately hydrogen, about 71% by mass, with 27% helium and all other chemical elements comprising the remaining 2%. This may sound surprising at first given that on our planet hydrogen and helium are not very common. Indeed, the two most common elements in the Earth are oxygen and silicon. We should remember though, that our planet comprises less than one thousandth of 1 percent of the mass in our solar system. Almost all of the mass in our solar system resides in the sun (nearly 99.9%), which is in fact, composed mainly of hydrogen and helium.

The density and temperature of the interstellar medium varies from place to place. Star formation occurs in the cooler denser regions because the low temperatures and higher densities favour gravitational contraction. This is because higher density implies more mass and hence more gravitational force, while lower temperatures imply lower relative velocities of the particles which means that they are more likely to remain in the region. (Temperature is just a measure of the average speed of the particles). These cooler denser regions are called dense clouds, which are themselves components of giant molecular clouds ('molecular' because the predominant form of hydrogen throughout them is in its molecular form, H_2).

Gravitational collapse is believed to be triggered by some external compression. For example, the collision of two different clouds, or by the impact of a shockwave from an exploding star (supernova). In this way, parts of the cloud becomes dense enough to become gravitationally unstable and contract. As it contracts it becomes denser to the point where the denser parts of the cloud each contract independently. Many of these fragments are the embryo from which stars are born. Clearly, this picture suggests that stars form together in stellar nurseries called 'star clusters', which is in agreement with the observations of young stars which indeed are usually found together in clusters. One such example is shown in **Figure 4.7**.

Let us now see what happens to a typical cloud fragment as it contracts. The gas molecules and dust fall inwards gaining speed and therefore temperature. In the process energy is released since the collisions of molecules lead to the emission of photons (light) which can initially escape the system. Inevitably though, the central region becomes so dense that the photons cannot quickly escape; they collide with the other molecules moving in a random direction with the effect that they only slowly diffuse outwards. Because energy cannot escape so quickly the temperature of the central region begins to rise rapidly. Eventually, after a few million years, the temperature of the central core becomes hot enough for nuclear fusion to occur – a temperature of about 10 million degrees Celsius. At this point a star is born.

Meanwhile, in another part of town, the material around the star has been evolving too. Not all of it falls into the central core because of its circular motion. As the particles orbit around the star, adjacent particles inevitably collide with each other, which flattens out the material into a disk and it also helps to make the orbits of

Figure 4.7: The Orion Nebula and Trapezium Cluster. The Trapezium cluster, marked by the four brightest stars, is a region of active star formation. (Credit: G. Schneider, K. Luhman et al, NICMOS IDT, NASA).

the particles approximately circular. **Figure 4.8** illustrates the flattening effects of the particle collisions.

The circulating disk of material will eventually form the planetary system. This picture can explain many of the observed features of our solar system, such as the fact that all of the planets orbit in approximately the same orbital plane in the same direction around the Sun.

Of course, we know somewhat less about the detailed formation of planets due to the difficulty of observing such systems, although there are some observations of young stars with discs of material

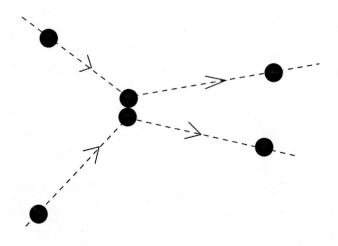

Figure 4.8: The flattening effect of particle collisions.

around them. The most plausible theory for the formation of planets sees large gaseous planets formed quite far from the star – typically greater than the distance between the Earth and the Sun. There are several reasons for this. First, the outer solar system encompasses a much larger volume and thus a larger amount of material should be present in the disk. Second, sufficiently low temperatures are required for ices of water (H_2O), methane (CH_4) and ammonia (NH_3) to condense out of the disk. These materials will eventually form the cores of the gas giants – after which hydrogen and helium gas can accumulate around the core. Clearly, the core must be large and dense enough so that its gravity can retain these light gases.

The necessity for icy materials to condense to form a dense core is simply because rock and iron are not abundant enough in the disk to form the cores of these gas giants. However, in the inner part of the solar system, the temperature is much higher implying that *only* rock and iron could condense to form planets. In our solar system the 'snow line' has been estimated to be just inside the orbit of

Jupiter, the first of the gas giants. In this way it is possible to understand why the inner 'terrestrial planets' – Mercury, Venus, Earth and
Mars are small rocky planets, while the more distant planets, Jupiter,
Saturn, Uranus and Neptune are much larger gas giants. Some properties of the planets in our solar system are given in the table below:

PLANET	Mass ($M_{Earth} = 1.0$)	Orbital Radius (AU)	Orbital Period (years)
Mercury	0.0553	0.387	0.2408
Venus	0.8150	0.7233	0.6152
Earth	1.000	1.000	1.000
Mars	0.1074	1.5237	1.881
Jupiter	317.9	5.20	11.86
Saturn	95.18	9.54	29.46
Uranus	14.54	19.19	84.01
Neptune	17.13	30.06	164.79
Pluto	0.0022	39.53	248.54

Table 4.2: The planets of our solar system.

While the basic properties of the planets in our solar system
seem to be more or less understood [*], the basic properties of many
extrasolar planets are very puzzling. We know that they do presently
exist very close to their star, but as I have discussed above, it seems
quite implausible for a large gas giant to form there. It is simply
too hot for the gases to condense to form a planet. Another problem is that the main components of the planet, hydrogen and helium
should have evaporated away. Basically in the high temperature en-
vironment next to the star the velocity of the hydrogen and helium
molecules could be large enough to allow them to escape into space.

---

[*]This is not to say that all of the observed properties of our solar system are
understood. There are still some puzzling aspects, such as, the orbital inclination
of Pluto (it orbits in a plane which is inclined at 17° to the plane of all the other
planets), also there are many puzzling features of various moons of the outer planets
etc. I will have more to say about some of these puzzles in chapter 6.

It is precisely this reason that the Earth's atmosphere has negligible hydrogen and helium and it is the reason why the Moon has no atmosphere at all. The gravity is not strong enough to retain these gases. However, for a gas giant these gases are the main components so it is obviously very important that the temperature and gravity are such that they do not significantly escape from the planet. While it can easily be checked that the present evaporation rate of a close-in extrasolar planet is negligible, this effect would be much greater in the past – when the planet was forming – since the radius of the planet was a factor of ten times larger than the present value. This means that the gravity at the surface was much smaller and simply not great enough to stop the gases evaporating into space. Thus, the planet would have simply boiled away and never have formed in the first place.

Nevertheless, we now know that large planets *do* presently exist right next to some stars. The Doppler shift and transit observations mean that this is an established fact. In view of these observations most theoreticians now argue that the planets must have formed far from their star – typically at a distance greater than about 5 AU – and migrated inwards through some 'planetary migration mechanism'. Such a thing may happen. After all, there are good reasons to believe that the Moon has migrated from a position much closer to the Earth to its present position, although the specific mechanism responsible for the Moon's migration could not work for large planets at large distances. I will return to the Moon in chapter 6.

Although planetary migration theories might be possible, there are also difficulties. First, it is necessary to explain why planets in a very large mass range, from about $0.5 M_J$ to at least $50 M_J$ can migrate towards their star. One has to also find a mechanism to stop the migration before the planet crashes into the star. Finally, one has to be able to explain strange systems such as the pair of resonant planets orbiting GJ876.

Precise Doppler measurements[25] of the star GJ876 have revealed two close-in planets with minimum masses of 0.56 and $1.89 M_J$. These two planets have orbital periods of 30.1 days and 61.0 days respectively. That is, the orbital periods are nearly in the ratio of 2:1. Such a situation is called an 'orbital resonance' and while common in very small bodies such as moons and asteroids is unprecedented among large planets. This poses major problems for migration scenarios because in most such scenarios the separation of the planets would be diverging as the inner planet should move inwards faster since it is closer to the star and moving faster relative to the outer planet.

Of course just about anything is possible – maybe even planetary migration. But a convincing specific theory has yet to be developed. The unconvincing nature of these theories is perhaps summed up best by one of the leading planet hunters, Geoffrey Marcy[26]:

> I think what's happening here is that we scientists are doing what we always do: hanging on to the paradigm until the last possible moment, the paradigm being that Jupiterlike planets form at 5 astronomical units (AU). If I had to put my money on the craps table, I would bet that we're going to learn later that giant planets can form closer in and that we've been deluded by our own solar system. But the theorists really hate that idea.

If the close-in extrasolar planets do actually form close to their stars, how can this be reconciled with the theory of planet formation? Recall that the main obstacle to close-in large planet formation is that the radiation from the host star would heat any ordinary matter to high temperatures. Mirror matter, on the contrary, would not be heated by the radiation from the star, so that it would remain much cooler. Perhaps this clue suggests that the weird properties of the extrasolar planets could be more simply explained if they are in fact made of mirror matter rather than ordinary matter...

Are close-in extrasolar planets mirror worlds?

Let us take another look at the formation of stars and planets, but this time imagine that the molecular cloud has a significant fraction of mirror matter. By 'significant fraction' I mean something like one of those simple fractions we all learned in primary school such as $1/10$, $1/2$, $3/4$...

Any random perturbation that triggered a gravitational collapse would occur quite independently for ordinary and mirror matter. Why? First, because they have separate electromagnetic interactions ordinary and mirror particles *do not* collide with each other. This means that the temperature (or mean velocity) of the ordinary and mirror matter components are quite independent. Regions which happen to be cool for ordinary matter might be hot for mirror matter or vice versa. This means that the gravitational collapse initially occurs in quite different places within the cloud for ordinary and mirror matter. Ordinary and mirror matter may also have different chemical compositions, that is, a different ratio of hydrogen to helium and different proportions of the heavier elements. They would also have different proportions of dust to gas. All these things depend on random perturbations and initial conditions which are quite different for ordinary and mirror matter.

In view of these differences, ordinary and mirror matter would collapse at quite different rates, as well as in different places. The end result is a natural segregation of ordinary and mirror matter in the formation of compact systems such as stars – meaning that solar systems containing predominately ordinary matter or predominately mirror matter would be formed. In other words, the necessary segregation of ordinary and mirror matter is actually expected, even if the original molecular cloud contained ordinary and mirror matter in equal proportions. [This simple but important observation was first made in 1982 by two Russian scientists, Sergei Blinnikov and Maxim Khlopov]. It is also reasonable to suppose that in some solar systems made of ordinary matter the amount of mirror matter could be of order a few percent (and vice versa). In such a system the mirror matter would collapse into the central region, in much the same way that the ordinary matter collapses producing a star. The result would be a planetary sized object near the star and in some cases the mirror matter would end up being swallowed by the star.

It therefore seems to be plausible that the final product of the gravitational collapse process in a large cloud containing both ordinary and mirror matter would be the formation of systems which feature a close-in mirror planet orbiting around an ordinary star. This may not happen often but it doesn't need to happen often. Only one out of every few hundred or so nearby stars are found to have close-in planets.

It seems reasonable to conclude that close-in extrasolar planets may be mirror worlds composed predominately of mirror matter. They do not migrate significantly, but actually formed close to the star, which is not a problem for mirror worlds because they are not significantly heated by the radiation from the star.

Are mirror worlds opaque?

While explaining very nicely the existence of close-in Jupiter sized extrasolar planets, there is also a conundrum. The problem is that a mirror world would be completely transparent. The ordinary photons from the host star would simply pass right through a mirror world because ordinary photons do not interact with mirror matter. This would not affect the Doppler shift measurements since this technique merely relies on the gravitational effect of the close-in planet to make the star move. Whether the planet is made of ordinary or mirror matter doesn't matter – the gravitational effect is the same and that's all that the Doppler shift measurements are sensitive to. However, in the case of the transit measurements things should be different. If the mirror planet is transparent, how can it make the star dimmer when it passes by? This effect has been clearly demonstrated for the system HD209458 as I have already discussed. In fact, I well recall that the news that HD209458 had an observable transit arrived just as my paper, pointing out that the extrasolar planets might be mirror worlds, was accepted for publication in the journal Physics Letters B. In that paper I suggested that no transit would be found in the mirror world hypothesis. But I was dead wrong.

Still, the interpretation of close-in extrasolar planets as mirror worlds seemed too nice to forget about. Maybe somehow they are opaque and do block out the light? But how? The first possibility which I had already mentioned in my published paper was due to the possible effect of a small transition force mixing ordinary photons with mirror photons. This same force will be discussed in more detail in connection with orthopositronium in the next chapter, so I will postpone a detailed discussion of it until then. Needless to say though, my colleagues Sasha Ignatiev and Ray Volkas and I, later showed (amongst other things) that this effect could not make the mirror world opaque. It can do various other things, but it can't make the mirror world opaque.

The solution to the problem turns out to be much simpler. While it *is* true that a mirror planet made of 100% mirror matter would be transparent such an idealised system cannot exist. There will always be some amount of ordinary matter which should have diffused towards the centre of the mirror planet when it was formed. The next step is to realise that the mirror planet is constantly being bombarded by protons and electrons coming from the host star. This out flow of gas is called the 'solar wind'. In our solar system it is responsible for the spectacular light displays observed at high northern and southern latitudes called the 'aurora borealis' and the 'aurora australis' (or the northern and southern 'lights' respectively). These lights are due to the collisions of the solar wind particles with the atoms in the Earth's upper atmosphere. The properties of the solar wind have been measured using various rocket and satellite experiments. It is found that the Sun is losing a small faction, $\sim 3 \times 10^{-14}$, of its mass per year. This may not sound like much, but over the lifetime of the solar system, which is a few billion years, it all adds up. Of course, only a fraction (about a percent or so) of the solar wind will strike the mirror planet.

What happens when these stellar protons and electrons from the host star hit the mirror planet? At the centre of the mirror planet there should be at least some non-zero amount of ordinary matter, which arises either as an initial component or due to the capture of some ordinary matter from interstellar space. The point is though, that the stellar protons will collide with any ordinary matter component in the mirror planet. These collisions will transfer some of the kinetic energy from the stellar protons to the ordinary matter component, which means that the stellar protons will lose energy and hence velocity, thereby getting themselves captured by the mirror planet. Therefore, any initial amount of ordinary matter in the mirror planet can act as a type of 'seed' which is fed by the solar wind from the host star.

The net effect is that during the lifetime of the solar system, the mirror planet would have accumulated a mass fraction of roughly 0.0003 (or 3 parts in 10,000) of ordinary matter. While it is only a small fraction, it nevertheless can have a very big effect on the opacity of the planet. Light from the host star which strikes this ordinary matter will be absorbed or rescattered. The small ordinary matter component can thereby make the planet opaque, but only in the region with a significant (column) density of ordinary matter. Evidently, to find out how much light gets blocked out during a transit we need to work out the distribution of ordinary matter within the planet.

The protons and electrons from the stellar wind will combine to form hydrogen gas which is heated by the ordinary radiation from the host star[*]. For close-in planets, the effect of this is to give the ordinary matter a 'surface temperature'[**] of between 1000 and 1500 degrees Celsius. That's very hot! How is this hot ordinary matter distributed within the mirror planet? The ordinary mass distribution depends mainly on just two things. First, there is the effect of gravity which is trying to force the ordinary matter into the centre of the planet. Opposing this, is the effects of pressure which, microscopically, arises from collisions of the particles. As gravity compresses the gas towards the centre, the pressure (or collision rate) increases until the two effects are balanced. In the technical literature this balancing act of nature is called 'hydrodynamic equilibrium'.

The compressing force of gravity depends on the amount of mass in the planet, which is dominated by the mirror matter since there is only a small ordinary matter component from the stellar wind. On the other hand, the opposing force of pressure results only from the ordinary matter component since ordinary and mirror

---

[*]Although the solar wind is composed primarily of protons and electrons, heavier ions such as helium, oxygen and sodium nuclei are also ejected from the star and should also be captured by the planet. This is important since it can affect the composition of the planets atmosphere which is potentially detectable during transit observations.

[**]Of course a planet made of gas technically doesn't have a real surface. 'Surface temperature' really means the effective temperature (or mean energy) of the thermally emitted photons (that is, the emitted heat).

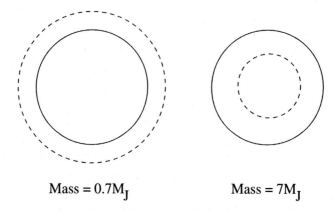

$$\text{Mass} = 0.7\text{M}_\text{J} \qquad\qquad \text{Mass} = 7\text{M}_\text{J}$$

Figure 4.9: A tale of two mirror matter planets, each with a small ordinary matter component. The one on the left has a total mass of $0.7M_{Jupiter}$ while the one on the right has a much greater total mass of $7M_{Jupiter}$. The volume occupied by the ordinary matter (dotted line) is very sensitive to the mass of the planet. The larger gravity of the heavier planet causes the ordinary matter to contract to a much smaller volume. While the mirror matter is also contracted by gravity, this effect is compensated because the heavier planet has 10 times more atoms to pack in. For this reason the radius of the mirror matter (solid line) is roughly the same for both examples.

particles do not collide with each other. Because the ordinary matter component, which is mainly hydrogen gas, is low in pressure and high in temperature, it can spread out over a large volume. In fact, the volume occupied by the ordinary matter component is roughly the same volume as occupied by the mirror matter – if the mass of the planet is the same as the mass of Jupiter. However, for larger masses the effect of gravity becomes stronger thereby causing the gas to contract into a smaller volume. This effect is illustrated in **Figure 4.9** (above).

Let us define the 'effective radius' of the planet to be the region within which the planet blocks out essentially all of the light from the host star. In the mirror world interpretation of the close-in extrasolar planets, this radius is just the extent of the region occupied by the ordinary matter component. This radius is, in general, distinct from the radius of the dominant mirror matter component. While the effective radius of mirror worlds depends sensitively on their mass, this is not true for ordinary (matter) gas giants. For example, if we compare Jupiter and Saturn, we see that Jupiter is approximately three times heavier than Saturn but it has a radius which is only about 18% larger. In this case, the contracting effects of the larger gravity of Jupiter is roughly balanced by the extra material that Jupiter has – it contains roughly three times more atoms than Saturn.

Clearly, if we can measure how the effective radius varies with mass we can potentially distinguish the mirror world interpretation from the conventional ordinary matter case. At the moment, there is only one extrasolar planet whose effective radius has been measured. This is the transiting planet orbiting the star HD209458. Recall that its mass is measured to be $0.69 M_J$ with an effective radius of about $1.3 R_J$ (where M_J, R_J are the mass and radius of Jupiter). It turns out though, that the large effective radius for this planet is consistent with the mirror world interpretation. It is also consistent with the planet being made of ordinary matter provided that it migrated to its close-in position *before* it was fully formed. What we really need is an accurate measurement of the size and mass of a second transiting planet which can be used to check the mirror world prediction that heavier planets should have a smaller effective radius[*]. The discovery of a second transiting system, which could happen at any time, may thus allow a means of distinguishing between the ordinary and mirror matter possibilities.

---

[*]For heavy planets ($> 2M_J$), the amount of light from the star which is blocked out during a transit could be expected to be very small $< 0.5\%$. It may therefore by possible that some of the heavy planets listed in table 4.1 are actually transiting systems. Searches for transits may have failed to notice the effect because their sensitivity was simply not great enough (recall the large error bars on Figure 4.4). Of course, the Hubble Space Telescope (HST) is much more sensitive (c.f. Figure 4.5) and its observations would surely register a transit if the orbital system had an edge-on orientation. For this reason it might be a good idea to check each close-in planet carefully with the HST to see if it is a transiting system.

Do ordinary and mirror planets orbit in different planes?

Already though, there are some interesting hints supporting the mirror world interpretation of the close-in extrasolar planets. One probable prediction of the mirror matter interpretation is that the mirror planets should orbit in a *different* orbital plane to the ordinary planets. This is because the collapse of mirror matter and ordinary matter are quite independent because of different random initial conditions. This prediction can be tested, and there is already some evidence that the close-in planets do sometimes occupy a different orbital plane to more distant planets.

The evidence comes from the star Upsilon Andromedae. Doppler shift measurements of the light from this star reveal that it has three orbiting heavy planets (see Table 4.1). From observations with the Hipparcos satellite – a space based telescope designed to measure positions of stars with unprecedented accuracy – a small stellar wobble has apparently been observed. This is an example of the astrometric technique discussed earlier. Anyway, these measurements allow an estimation of the inclination angle of the orbital plane of Upsilon Andromedae's most distant orbiting planet, which was roughly estimated to be[27] $\sin \theta_I \approx 0.4$.

On the other hand, if all three planets are co-aligned in the same plane then it has been argued[28] that the system may *not* be stable unless $\sin \theta_I \gtrsim 0.75$. The reason that stability gives a lower limit on $\sin \theta_I$ is simply because the larger $\sin \theta_I$ the smaller the mass of the planets (since Doppler shift measurements have already determined the product of $\sin \theta_I \times Mass$). Obviously, if the masses of the planets are smaller then so are the gravitational perturbations, thereby leading to more stable orbits.

Searches for reflected light

Another interesting hint which supports the mirror matter interpretation comes from searches for reflected light from the close-in planets. When light from the host star strikes a planet, a fraction of it is reflected back into space with the rest being absorbed thereby heating the planet. The absorbed energy is eventually released at much longer wavelengths as heat energy, mainly in the infrared frequency region. In principle, both reflected light and the infrared heat emission could be searched for. At the present time, measuring reflected light is somewhat easier and this reflected light has been searched for by several groups.

Reflected light is distinguished from stellar light by looking for a secondary component that varies in brightness as the planet moves around the star. The maximum reflected light occurs when the planet is in opposition, that is behind the star when viewed from Earth (assuming of course that the orbit is not exactly edge-on, so that the reflected light is not blocked by the star). The reflected light also varies in frequency due to the large orbital velocity of the planet (which can be distinguished from the Doppler shift of the stellar light because the planet's orbital velocity is much greater than the star's orbital velocity). The search for reflected light is made easier if the apparent brightness of the star is large and also if the planet is nice and close to the star. Both of these features help to increase the observable reflected light signal.

Out of all of the extrasolar planets discovered to date, the most suitable system for a reflected light search is the planet orbiting around the star Tau Bootis. Tau Bootis has a close orbiting planet with orbital radius of about seven million kilometres. Several groups of astronomers have searched for a reflected light signal from Tau Bootis, but so far no reflected light has been detected within the sensitivity of the observations. This non-detection allows us to place a maximum value for the fraction of light reflected by the planet orbiting Tau Bootis. The fraction of reflected light is called the 'albedo', and I use the symbol p for it. [$p = 1$ corresponds to 100% reflected light and 0% absorbed, while $p = 0$ means no reflected light and 100% absorbed].

To translate the non detection of reflected light into a maximum value for p we need to know the radius of the planet. The radius of the planet can be estimated from its mass. The planet orbiting Tau Bootis is quite heavy with an estimated mass of about seven times the mass of Jupiter. That is, the planet around Tau Bootis is about 10 times heavier than the transiting planet orbiting HD209458. If Tau Bootis were made of ordinary matter then its radius has been estimated to be about 20% larger than that of Jupiter. In which case, the non-detection of reflected light implies[29] a maximum value for p of just 0.22. In other words, if Tau Bootis is made from ordinary matter it must reflect less than 22% of the light falling on it – which is really quite a stringent limit. By comparison, it is known that Jupiter reflects about 55% of the Sun's light.

Alternatively, if Tau Bootis is a mirror world, its large mass implies a large gravitational force. This means that the ordinary matter which is captured from the solar wind of the host star would be much more tightly compressed. The effective radius of this ordinary matter would only be about half of the radius of Jupiter (c.f. figure 4.9). Because of its small size the amount of reflected light is much smaller. In this case, the non-detection of reflected light implies a very weak maximum value for the albedo, $p < 1.3$. Actually, this is no limit at all since we know that p must be less than one. No planet can reflect more light than the amount of light falling on it. (Although it is possible for a planet to produce energy from contraction and radioactivity, such energy sources cause emission in the infrared region, which is a different frequency region to reflected light searches). Evidently, the mirror world interpretation of the close-in planet orbiting Tau Bootis automatically explains the non-detection of reflected light. Put another way, if people had observed reflected light for Tau Bootis then the mirror world explanation for it would have been excluded.

In summary, the mirror matter interpretation of the close-in extrasolar planets seems to be quite successful in explaining their observed properties. Further observations, especially of more transiting systems, should be able to provide a clear means of distinguishing between the ordinary matter and mirror matter interpretations. In the meantime, I would argue (on most days) that this mirror matter interpretation of the close-in planets seems, on balance, more likely IF mirror matter exists. But this looks like a big if!

If these close-in extrasolar planets are manifestations of the mirror world, then it seems very natural that the 'dynamical mirror image system', consisting of a mirror star with an ordinary planet, should also exist. Such a system would appear to ordinary observers as an 'isolated' ordinary planet. That is, a planet apparently free-floating, not associated with any visible star, since the mirror star would be invisible. Remarkably, such apparently 'isolated' planets have recently been identified in the σ Orionis star cluster and several other nearby star clusters...

Observation of 'isolated' planets

One might think that small objects such as isolated planets would be very faint – and this is certainly true. Nevertheless, if they are young then they should still be quite hot. A hot fire is much more luminous than a cold fire and so it is with planets. So, the place to seek faint light-weight stars or planetary mass objects is in the region where they are born.

As I mentioned earlier, stars form by gravitational collapse within large clouds of gas and dust. I was going to write that stars come in various shapes and sizes, but this isn't really true. The shape of all stars is basically spherical, although they can be flattened a little by their own rotation. What concerns us now, though, is not stellar shape but the weight or mass of stars. Stars, when they are born, have a large range of possible masses. This *mass distribution* is described in the technical literature by something called 'the initial mass function'. This quantity is just the probability of producing a star of a certain mass, or rather, in a certain small mass range. The theory of gravitationally collapsing gases within molecular clouds allows people to estimate the shape of the initial mass function. This is plotted in **Figure 4.10**[*].

As the figure shows, the most likely mass for a star is around two thirds of the mass of the Sun. For masses greater than about $0.1 M_{sun}$ the expected function agrees roughly with the observations. If the mass of the star is less than about $0.1\ M_{sun}$ it does not get hot enough to burn hydrogen. This means that such stars are very faint objects which makes them difficult to detect. Stars lighter than $0.1 M_{sun}$, but heavier than about 13 times the mass of

---

[*]Technically, the initial mass function describes the number of stars per unit area of the Milky Way's disk per unit interval of logarithmic mass.

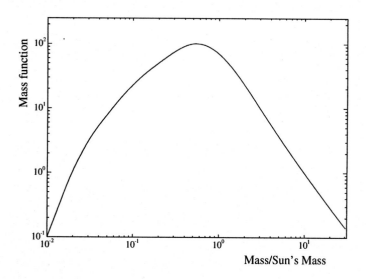

Figure 4.10: Expected initial mass distribution of stars in our galaxy, called 'the initial mass function'.

Jupiter[*] can briefly burn deuterium. These objects are called brown dwarfs. Observations have shown that such low mass objects are relatively rare in star clusters. The first example of a brown dwarf was discovered only recently in 1995.

As one goes down in mass even further one would obtain planetary mass objects. Note though that the mass function (plotted in figure 4.10) refers to *isolated* objects; planets orbiting stars do not count in the mass function. However, given the scarcity of brown dwarfs there was a strong expectation that lighter isolated objects of planetary mass would be extremely rare. Theoretically this was also expected, as indicated in Figure 4.10. Such low mass objects would originate from such a small cloud fragment that the force of gravity would be too weak to initiate the gravitational collapse. In general, one needs enough gravity to overcome the thermal energy of the particles in the gas otherwise they simply escape rather than

---

[*]The mass of Jupiter is roughly 1/1000 that of the Sun, which means that $13 M_{Jupiter} \approx 0.01 M_{sun}$.

being captured within the fragment.

While theoretically unexpected they may be – nevertheless somebody forgot to tell the star clusters about the theory... In yet another unexpected recent discovery, apparently isolated low mass objects of planetary mass are in fact very common! First reports of such objects appeared in 1998 when a Japanese group led by Motohide Tamura of the National Astronomical Observatory of Japan discovered two extremely dim bodies in the Chamaeleon star cluster. They estimated that the mass of these objects was about ten times the mass of Jupiter. Independently, two British astronomers, Phillip Lucas of the University of Hertfordshire and Patrick Roche of the University of Oxford have discovered 13 planetary mass objects in Orion's Trapezium star cluster – the same star cluster that was shown in Figure 4.7.

Such was the unexpected nature of these discoveries that the initial reaction of other workers in the field was highly sceptical. It was initially thought by many that the conclusions of Lucas and Roche were somehow mistaken. For example, Lynne Hillenbrand, an astronomer at Caltech, suggested that the so-called 'isolated planets' were more likely distant stars outside the cluster masquerading as giant planets[30]: 'I'm content with the conclusion that they are unrelated objects'

The confirmation that such objects really are planetary mass objects came from a third group lead by the Spanish astronomer Maria Zapatero Osorio of the Institute of Astrophysics in the Canary Islands[31]. Using visible and infra-red light detecting sensors on telescopes in Spain, the Canary Islands, and Hawaii, they discovered 18 objects which have the right broad features to be planetary objects in the σ Orionis star cluster. Three of these objects were then studied in detail. The infra-red and optical spectrum of the emitted light from these objects was carefully analysed in the Keck telescope in Hawaii. These spectrum measurements confirmed that the new objects were cool, with planet-like temperatures, and matched closely their expectations. The spectra showed that the bodies were not distant stars or far-off galaxies masquerading as planet-like objects.

These planets have an estimated mass of $5-15 M_{Jupiter}$. Planets lighter than this mass range would be too faint to have been detected,

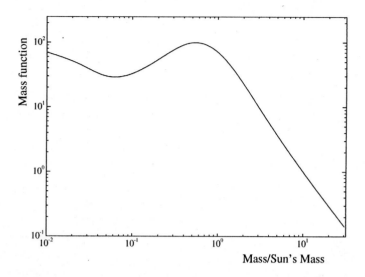

Figure 4.11: Mass distribution of stars in our galaxy as in Figure 4.10, but extended to include the observations of numerous low mass objects in the σ Orionis cluster.

but will be observable in the future if they exist. The observed planets appear to be gas giants which do not seem to be associated with any visible star. Given that the σ Orionis cluster is estimated to be less than 5 million years old, the formation of these 'isolated' planets must have occurred within this time. This means that they can't orbit faint stellar bodies such as white dwarfs or exotic objects such as black holes, both of which take much longer to form and/or would have other tell-tale signs. If they really are isolated objects then the mass plot of Figure 4.10 must be modified as shown above in **Figure 4.11**.

Such a behaviour though, not only looks strange, it is strange. It is completely unexpected on the basis of current theories. As I indicated earlier, the standard theory of star formation is unable to explain the existence of numerous isolated planetary mass objects since they are too small to initiate gravitational collapse within clouds. Can mirror matter help?

Mirror matter comes to the rescue again

The first suggestive clue is that the mass of these isolated planets is roughly the same as some of the close-in planets observed around nearby stars. Maybe there is a connection. Recall, I suggested earlier that close-in ordinary planets could be more naturally interpreted as mirror worlds orbiting ordinary stars. If such systems exist, surely the opposite type of system, with ordinary worlds orbiting mirror stars, should also exist. If it did, it would appear to us as an 'isolated' planet because the mirror star would be unobservable. Mirror matter emits only mirror light which is undetectable to us ordinary people. Thus, the 'isolated' planets may just be ordinary planets orbiting mirror stars. They appear 'isolated' because the mirror stars are invisible. With this interpretation of the 'isolated' planets Figure 4.11 can be decomposed in terms of the two distinct populations as given in **Figure 4.12**.

There is a very simple way of testing this idea. Recall that the orbital motion of stars, resulting from the gravitational tug of their planets, has been detected by measuring the frequency change of the light (the Doppler or 'fire engine' effect). Clearly, if the 'isolated' planets really orbit invisible mirror stars then their light must contain a periodic frequency shift too. In fact, the velocity of an orbiting planet is typically several hundred times greater than the velocity of the star, which should make this effect quite easy to detect despite the low brightness of the planets. This effect will be especially large for close-in ordinary planets where the velocity is larger and the orbital period very short. If the orbit of the planet is more distant then it could be possible to directly observe its orbital motion (astrometric technique), but this may take several years of observations. The important thing is that this idea that the isolated planets are in fact orbiting invisible mirror stars has distinctive observational tests. It can be proven true or false.

I have now explained the third specific evidence for the existence of mirror matter. Clearly, these three astrophysical puzzles are somewhat interconnected and mutually supportive. For example, if close-in extrasolar planets are made of mirror matter then it would

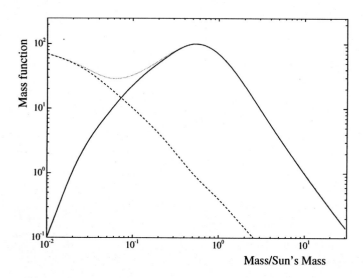

Figure 4.12: Mass distribution of stars in our galaxy in the alternative interpretation of 'isolated' planets as ordinary planets orbiting mirror stars. The solid line is the 'true' mass function of isolated ordinary stars while the dashed line is the number of mirror stars with ordinary planets orbiting around them. The dotted line is the sum of the two components, which is the measured mass function because the mirror stars are not observed.

be surprising if the dynamically opposite system comprising a mirror star with ordinary planet did not exist. Or, if isolated planets are indeed ordinary planets orbiting mirror stars then it is plausible that MACHOs (discussed in chapter 3) should exist and are just mirror stars. These 'three wonders of the mirror world' are summarized in **Figure 4.13**.

Discussion

It is known that ordinary matter produces an extremely wide range of structures in the Universe. There is a large range of galaxies in all shapes and sizes. For example, there are small dwarf

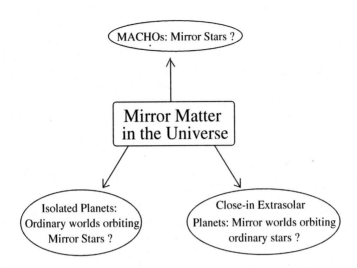

Figure 4.13: Three wonders of the mirror world.

spheroidal galaxies with estimated masses of roughly $10^7 M_\odot$ ($M_\odot = M_{sun}$), elliptical galaxies with masses ranging from $10^8 M_\odot$ to $10^{13} M_\odot$, spiral galaxies with masses $10^9 M_\odot$ to $10^{12} M_\odot$. As the nomenclature suggests these galaxy types are distinguished by their shapes, and also within each class there are many differences, such as different mass to light ratios. Within galaxies there are smaller structures called 'globular clusters' which typically contain about 10^5 stars and there are a large variety of stars too...

The observational evidence also suggests that dark matter is equally diverse. It exists in dwarf galaxies, in spiral galaxies such as our Milky Way galaxy and in other even larger galaxies. There is also evidence from gravitational lensing that there may even be large invisible galaxies containing almost no ordinary matter at all. Furthermore, we have already seen in chapter 3 that there is evidence that dark matter forms compact systems like stars, yet are invisible (MACHO observations). Also, we have seen in this chapter that it is possible to interpret close-in extrasolar planets and 'isolated' planets as dark matter manifestations.

The observations evidently suggest that dark matter is as complex as ordinary matter. This feature is evidence supporting mirror matter, since mirror matter should form equally complex structures as does ordinary matter. The essential microscopic similarity of ordinary and mirror matter will ensure this. The microscopic symmetry between the interactions of the ordinary particles and the interactions of the mirror particles will inevitably imply some macroscopic similarities. Of course, we would like to have a precise understanding of all the observed properties of the Universe. It would be nice to be able to predict, for example, the population of systems with mirror planets orbiting ordinary stars and vice versa. Predicting the behaviour of complex systems of particles is much like predicting the weather, which is of course, a problem of the evolution of a gas containing a large number of particles. It is very difficult. In fact, it is even more difficult than the weather because of the lack of knowledge of the initial conditions. The chemical composition, the temperature distribution, the relevant amounts of ordinary and mirror atoms are all unknown.

Naturally, one thing leads to another, and so it is with initial conditions. The problem can be traced to the early history of the Universe. Within the framework of the standard big bang model, the Universe was assumed to begin a finite time ago, it's now about 10-15 billion years old. Starting from a gigantic explosion the Universe was created and has been expanding ever since. Understanding this early history of the Universe is very difficult due to the paucity of information available to us. Paucity of information has, rightly or wrongly, never stopped people from extrapolating and speculating, but we must at least recognise the limitations of such endeavours. I will not dwell upon it, but merely point out one relevant point. An important unknown quantity is the 'initial conditions' affecting the early evolution of the Universe. *Exact microscopic symmetry does not imply identical initial conditions for the ordinary and mirror matter.* For example, we cannot automatically assume that the initial number of ordinary and mirror nucleons are equal, or that their temperature distribution is the same. This could be hypothesised in addition to the hypothesis of microscopic mirror symmetry in particle interactions, but it doesn't logically follow from the hypothesis of mirror symmetry itself[*]. In any case, the point is that these unknowns make the already difficult task of understanding the early evolution and generation of large scale structure in the Universe even more difficult. At the present time there is no completely satisfactory description of the formation of structures such as galaxies, which does suggest things are complicated, possibly by the existence of mirror matter.

Perhaps most important for the existence of mirror matter in the Universe is that it can plausibly explain the three *specific* sets of observations already discussed in this and the previous chapter: invisible star sized objects in the halo of our galaxy (MACHOs), close-in heavy planets around nearby stars, and the existence of numerous 'isolated' planets. If these three things are mirror world manifestations then this will become clearer in the near future. In particular the inevitable discovery of a second transiting system (which may happen at any time and several new transiting systems could be expected to be discovered in the next five years) could greatly clarify the nature of the close-in heavy planets around nearby stars.

---

[*]In fact, a Universe with equal numbers of ordinary and mirror atoms (or more precisely, nucleons) appears to be disfavoured. The totality of observations seem to indicate that the exotic component of the dark matter is, of order, 10 times the amount of ordinary matter. This suggests that we would need 10 times more mirror atoms than ordinary ones. This asymmetry in the amount of ordinary and mirror matter in the Universe may be due to either *asymmetric* initial conditions or it may be that the early evolution of the Universe produced this asymmetry itself via random evolutionary effects.

*The sky split apart and a
great fire appeared. It became
so hot that one couldn't stand it.*

An eyewitness 90 kilometres from
the Tunguska explosion of 1908.

PART III

Evidence for Mirror Matter in the Laboratory and Solar System

Chapter 5

Mirror Matter and Positronium

In part II we have seen that astronomical observations have revealed tantalizing evidence that mirror matter really exists in the Universe. Despite our inability to see it, we can (tentatively) infer its existence by its gravitational effects. Actually, this is not the only place to search for mirror matter. If it exists its effects will be felt on the Earth as well as in the Heavens. Such terrestrial effects are of crucial importance since they allow the idea to be put under 'the microscope', thereby providing a means of potentially rigorously proving that mirror particles really exist.

Mirror matter on Earth and in the Heavens

There are two broadly different ways in which mirror matter can show itself. One is its implications for some of the largest objects known: planets, stars and galaxies. Mirror matter is important on these very large scales because it is influenced by gravity. What about the influence of mirror matter on microscopic particle processes? I have already explained that mirror matter and ordinary matter do not interact with each other by any of the *known* forces, except via gravity. Is it possible for *new* types of interactions to connect ordinary and mirror matter? Certainly such interactions would have to be very small if they exist, but the question is – can they exist? The answer is yes – but in only two ways. In this and subsequent chapters I will describe these new types of interactions and explain how their effects may be observed. In fact, there is evidence that the observable effects have already been observed!

I have discussed earlier that the interactions of sub-atomic particles seem to obey various symmetries such as rotational symmetry, translational symmetry and more abstract symmetries called Lorentz and gauge symmetry. The gauge symmetries govern the three non-gravitational forces: electromagnetism, strong and weak nuclear forces, while gravity can be related to the curvature of space-time. This distinction between the gravitational and non-gravitational forces is the reason why we expect both ordinary and mirror particles to experience the same gravitational force. Ordinary and mirror particles live in the *same* space-time so they are both affected by the curvature of space-time. For example, the motion of an ordinary planet orbiting around a star does not depend on whether the star is made of ordinary matter or mirror matter because the curvature of space-time is simply due to the star's mass. However, the three non-gravitational forces are related to symmetries on an 'internal' space which is nothing to do with space-time. Nobody has ever seen this internal space, so it is not known how big it is or how many symmetries it contains. For this reason it is possible for ordinary and mirror particles to have independent but otherwise identical gauge symmetries. The effect of this is that the non-gravitational forces can act separately on ordinary and mirror particles.

Given these symmetries, the only way in which ordinary and mirror matter can interact with each other besides gravity is through 'transition forces' mixing ordinary with mirror particles. This type of interaction can only occur for neutral particles, since the conservation of ordinary electric charge forbids, for example, an electron from becoming a mirror electron. Ordinary and mirror particles have independent gauge symmetries which means that mirror particles have no ordinary electric charge. (However they do have mirror electric charge). The only known fundamental neutral particles are photons and neutrinos. One might also add the neutron, but actually the neutron (and proton) are not believed to be really elementary. Indeed, as I already mentioned in chapter 2, neutrons and protons are believed to be composed of quarks. The only thing that we really need to know about all this is that there aren't any neutral quarks, so there can't be any quark-mirror quark transition forces. Thus, it seems that symmetry arguments do not forbid transition forces between photons and mirror photons and also between neutrinos and mirror neutrinos. One thing that we have learned from the study of particle interactions is that if something is not forbidden then it usually occurs, and so it is worth taking such effects seriously.

To summarize, there are only two ways for the (known) ordinary and mirror particles to interact with each other besides gravity. That is, by a photon-mirror photon transition force (also called 'mixing' force), and by neutrino-mirror neutrino mixing. The effect of neutrino-mirror neutrino mixing and the substantial evidence for it will be described later on. Our purpose now is to describe the effects of photon-mirror photon mixing.

Photon-Mirror Photon mixing

What effect does photon-mirror photon transitions have? As I discussed in Chapter 2, microscopic interactions can be discussed in terms of interaction diagrams. The photon-mirror photon transition force is represented by the interaction diagram shown in **Figure 5.1**.

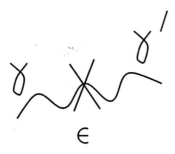

Figure 5.1: The photon-mirror photon transition force is simply represented by a 'cross' interaction in which a (virtual) photon (γ) turns into a (virtual) mirror photon (γ'). The parameter ϵ characterises the strength of the transition force, in much the same way in which the electric charge (e) characterises the strength of electromagnetism.

Recall that the photon is the force particle for electromagnetism. The electromagnetic force arises from the interactions of the charged elementary particles, electrons and protons with each other through the exchange of an ordinary (virtual) photon. In the case of electrons this was illustrated in Figure 2.2.

In the absence of any photon-mirror photon mixing, an ordinary electron cannot interact with a mirror electron because ordinary photons do not interact with mirror electrons (and mirror photons do not interact with ordinary electrons). However, if there is a photon-mirror photon transition force, then now an ordinary electron can actually interact with a mirror electron. What happens is that the ordinary electron can emit a photon which then undergoes a transition to a mirror photon which then interacts with the mirror electron. This effect is illustrated in the interaction diagram given in **Figure 5.2**. The net effect of the transition force is to make mirror electrons interact slightly with ordinary electrons. *That is, mirror electrons behave as if they have a tiny ordinary electric charge.* The size of the effect depends on the strength of the photon-mirror photon transition force – which is characterised by the parameter ϵ. If I use the symbol e for the ordinary charge of an electron, then the mirror electrons effectively have an ordinary electric charge of ϵe (and $\epsilon \ll 1$). [A similar effect also happens for mirror protons too].

This is really very important if it exists. It means that ordinary and mirror matter can repel (or attract) each other. In chapters 1,2 the merits of trying to pick up a rock made of mirror matter was discussed. It was suggested that a mirror rock would fall through our hand and the Earth as well, because it was assumed that the only force acting between ordinary and mirror matter was gravity. In the introductory discussion I neglected the possibility that an ordinary photon-mirror photon transition force could exist. If this force does exist then a rock made of mirror matter may not fall through our hand when we pick it up. The point is that the force of gravity is not very strong and even a small electromagnetic coupling can be enough to oppose the feeble gravitational force. An important question to answer is to find out what the strength of this new force is. That is, how big is ϵ?

Clearly, if the force connecting ordinary and mirror matter was

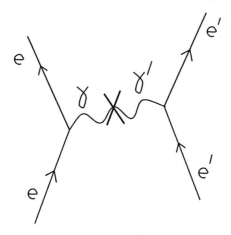

Figure 5.2: If the photon-mirror photon transition force exists then an electron can interact with a mirror electron.

big enough then its effects would have already shown up in laboratory experiments. An obvious place to start is by looking at experiments searching for exotic particles with tiny ordinary electric charges. There have been such experiments, and the most sensitive laboratory limit on these 'milli charged particles' implies that[32] $\epsilon < 0.0001$. However, this is not the most sensitive way to search for the mirror world effect. Sometimes things are more subtle. The question that needs to be asked in this type of situation is the following. Out of all the experiments that have been done in the history of the world, which experiment would be most sensitive to photon-mirror photon mixing? The answer it turns out is the effects on a system called 'orthopositronium' – a weird type of atom composed of matter and anti-matter. Before diving into the deep end we need to briefly discuss some more things about atoms.

More about atoms

The simplest stable atom is hydrogen. Hydrogen consists of an electron and a proton which are bound together by the electromagnetic force. The stability and other properties of hydrogen was once a mystery – back in the good old days in 1911 it was at the forefront of modern research. The electron had been discovered in 1897-1899 by J. Thomson and the proton had just been deduced to exist by Rutherford. It then seemed that atoms were made of electrons and protons (and lots of empty space). But how? According to 19th century ideas, hydrogen should not exist at all. An electron moving around a proton should radiate light and the electron should spiral into the proton.

There has been much progress since 1911 and today we think we understand hydrogen very well indeed. As (almost) no mechanic will tell you, the stability of hydrogen is explained by quantum mechanics. Our ordinary notions of the behaviour of particles are built up by observing large objects moving over large distances. The notions of forces and particle trajectories that were developed by Newton and others simply do not work for atoms. Therefore, for very small distances, such as the behaviour of an electron within an atom, radical revision is required. Particles, it turns out, behave in a very strange way. On small distance scales, they tend to have the attributes of waves. Wave-like phenomena such as diffraction and interference have all been observed to occur for particles such as electrons and neutrons. Thus, the experimental discovery of electrons and protons eventually led to a new understanding of microscopic phenomena, which is called 'quantum mechanics'. This theory encompasses the Newtonian or classical picture when distances are large, but is radically different on atomic distance sizes. With this theory the hydrogen atom became very well understood and the theory has had a glorious history ever since, with many successful predictions and developments. Today, it is the backbone of all undergraduate university courses in physics.

Clearly, by learning about the behaviour of particles on atomic distances, great progress was made in understanding the stability and other properties of hydrogen (and indeed the other chemical elements). The stability of hydrogen also depends on the properties of the particle interactions themselves. For example, if the neutron happened to be lighter than the proton, then hydrogen would be unstable. Hydrogen would decay into a neutron and a neutrino via the weak nuclear force,

$$\text{Hydrogen} \rightarrow n + \nu$$

and that would be bad news for us. But the neutron is not lighter than the proton, it is 0.14% heavier so we are in luck – we can exist. Or maybe, because we exist the neutron is not lighter than the proton. Perhaps there are alternative Universes where the neutron is lighter than the proton, but there are no people to wonder why, or maybe not. Such uncertain philosophical debates go by the name of the 'anthropic principle'.

Returning now to our Universe, we know that the neutron is heavier than the proton. This means that hydrogen cannot decay via the weak interaction because of the conservation of energy. Lighter particles cannot decay into heavier ones ($E = mc^2$ and all that). But why should energy be conserved? Good question – but there is a good answer: symmetry. I was going to add "In addition to the fact that no process has ever been observed to violate energy conservation" but this isn't really true. When β-decay was first observed it appeared to violate energy conservation. In β^--decay, neutrons in certain nuclei are spontaneously transformed into protons with an observed electron emitted. The fundamental process involved appeared to be

$$n \rightarrow p + e$$

Taking account of the original energy of the nuclei, then conservation of energy tells us that the energy of the decay products should simply add up to the original energy. But this is not what was observed. The total energy of the decay products was observed to be *less* than the original energy of the nuclei before it decayed.

Energy conservation is of course very important – we should all try to save energy. While saving energy is important for the present and especially the future – physicists did their bit by saving the energy conservation law. Instead of concluding that energy conservation was violated, physicists (some of the them at least) concluded that an invisible particle – called a neutrino was being emitted along with the electron. In this way it could be argued that energy was really conserved, it only *appeared* to be not conserved, because the energy being carried by the neutrino wasn't detected.

Initially, not all people believed in the neutrino. Some, including luminaries such as Niels Bohr, thought it more likely that energy was simply not conserved. However, within a few years it was shown that this seemingly bold invisible neutrino assumption could explain many of the observed properties of β-decay such as the detailed spectrum of energies of the observed electrons. By the late 1930's the existence of the neutrino was widely believed, despite the fact that nobody had ever seen one. Even the sceptics had become believers. Physicists' faith in the existence of the neutrino was eventually rewarded when the neutrino was finally discovered in the 1950's.

In a certain sense though, the discovery of the neutrino was surprising since it was thought by many that it was so elusive it might never be directly observed. However, with the advent of nuclear reactors, the ability to produce huge numbers of them became possible, leading to their eventual discovery. Since that time, remarkable technological advances have been made, and neutrinos produced from the Sun as well as a by-product of cosmic ray interactions with the atmosphere, are routinely detected. Furthermore, as I discussed in chapter 3, even neutrinos from a distant supernova explosion were observed in 1987. I will talk more about my friends the neutrinos in chapters 7,8.

What has energy conservation got to do with symmetries? In 1918 Emmy Noether made a very important observation. She showed that for *every* continuous symmetry of the particle interactions, such as translational symmetry or rotational symmetry, there was necessarily a conserved quantity. Noether showed that conservation of energy is related to time translational symmetry, conservation of momentum to space translational symmetry. Rotational symmetry leads to conservation of something called 'angular momentum', gauge invariance to electric charge conservation and so on.... Hence, if energy were not conserved the laws of physics would have to change in time – which might make it hard to nail them down. This symmetry reason is the main reason why it is thought that energy is always conserved. Symmetry is so elegant!

The upshot of all this is that the hydrogen atom is stable. There are many other stable atoms and this is all more or less understood. Another completely different type of atom can be made from anti-matter. Anti-hydrogen is made from positrons and anti-protons. Anti-hydrogen is very similar to ordinary hydrogen. It is also stable (in isolation or with other anti-atoms), however when anti-matter meets ordinary matter, both matter and anti-matter annihilates into pure light energy. There is yet another known type of atom which is a mixed form of matter and anti-matter. One can imagine building an atom from an electron and a positron. Actually we need not just imagine it, it has already been done. This type of atom is called 'positronium'. There are also many other matter - anti-matter atoms such as quarkonium (made of a quark and an anti-quark), muonium (made of a muon and anti-muon) and of course, pandamonium. The last system is perhaps most familiar to people who visit the Zoo...

What is orthopositronium?

Electrons and positrons are each stable particles. If left on their own they just sit there quite happily. However, when they are together in a bound system called positronium, at any given time interval they have some chance of meeting each other and annihilating into photons. The end result is that positronium is unstable. If the spins of the electron and positron are aligned so that the system has total spin 1, then the system is called 'orthopositronium'. If it is anti-aligned so that the total spin is 0 then the system is called 'parapositronium'. The decay properties of orthopositronium are quite different to parapositronium. Orthopositronium decays into three photons, while parapositronium decays into two photons. The reason for this is well understood, it is due to conservation of angular momentum and the quantum mechanics of identical particles.

Now the force governing orthopositronium is electromagnetism. These days electromagnetism is also very well understood. This means that the mean time it takes for orthopositronium to decay can be calculated very precisely in terms of the mass of the electron and its electric charge – both of which are known with extremely high precision. While the orthopositronium lifetime *can* be calculated very precisely, the calculation is not particularly easy. In fact, accurate calculations of the properties of such bound state systems are quite difficult and technical. Fortunately there are experts who specialize on doing exactly these types of calculations. The result of these calculations is the following orthopositronium lifetime:

$$\tau_{oPs}(theory) = \left(1.420805 \times 10^{-7} + \Delta\right) \text{ seconds.}$$

The term Δ represents a tricky higher order correction. For a long time this term remained unknown, but on general grounds it was expected to be very small. In a timely publication (published in April 2000) three experts in tricky calculations finally calculated the Δ term. G. Adkins, R. N. Fell and J. Sapirstein showed that it was indeed a very small correction, thereby obtaining the following accurate prediction of the orthopositronium lifetime[33]

$$\tau_{oPs}(theory) = 1.42047 \times 10^{-7} \text{ seconds.}$$

Of course, the above calculation gives the orthopositronium lifetime in the absence of any new physics such as the effect of mirror matter, which I will now illuminate.

As a microscopic (or quantum) theory, the electromagnetic force arises from the exchange of a 'force particle'. The force particle for the electromagnetic force is the photon, that is the particle that makes up light. In the case of hydrogen, the electron and proton continually exchange a photon which generates an effective potential energy. This effect comes from the interaction diagram shown in **Figure 5.3**.

For the hydrogen atom the effective potential energy can be calculated very precisely indeed. It is -13.5984 eV. The negative sign occurs for systems that are bound. It means that the total energy of

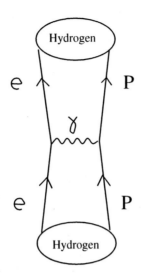

Figure 5.3: Hydrogen is composed of a single proton (P) and an electron (*e*) which is bound together by the exchange of photons (γ). Photons are the force particles of the electromagnetic force.

the hydrogen atom is actually less than the energy of the electron and proton separately. If we were to weigh the hydrogen atom, its weight would be less than the weight of its parts, because of mass-energy equivalence ($E = mc^2$ and all that).

In the same way, the effective potential can be calculated for positronium, and I will consider the interesting special case of or-thopositronium. The exchange of a virtual photon also generates an effective potential in several different ways, as shown in **Figure 5.4**. The second diagram of that figure is of most significance to us and only occurs for matter-antimatter atoms such as orthopositronium. In fact, it is the reason why photon-mirror photon transitions can affect orthopositronium and not ordinary atoms such as hydrogen.

The realisation that orthopositronium is particularly sensitive to the existence of mirror matter was first recognised by Shelly Glashow in 1986. Glashow, despite winning the Nobel Prize in 1979 is in fact a first rate scientist! Perhaps even a giant of 20th century

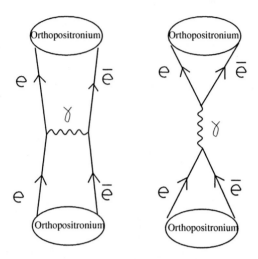

Figure 5.4: Orthopositronium is made of an electron (e) and positron ($\bar{e}$) bound together by exchanging a photon. The photons can be exchanged in two different ways. In the first way (the diagram on the left) an electron emits a photon which is absorbed by the positron (or vice versa). Alternatively (the diagram on the right) an electron and positron can come together and annihilate into a single photon which can again produce the orthopositronium state.

physics – being more than 185 cm tall! Fortunately he also has a great sense of humour. In his autobiography he states[34] 'Clearly 1932 was another great year for physics. It was the year I was born...'. Perhaps 1932 was indeed a good year. Let us now see what it is that photon-mirror photon transitions do to orthopositronium.

What orthopositronium did next

If photon-mirror photon mixing exists then there is another type of potential energy. A sort of 'off diagonal' energy which mixes an orthopositronium with its mirror partner. (See **Figure 5.5**). This type of mixing means that neither orthopositronium nor mirror

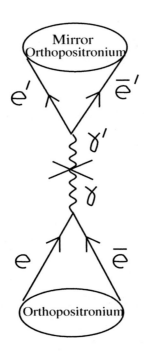

Figure 5.5: If photon-mirror photon transitions exist (indicated by the cross on the diagram) then there is a 'potential energy' mixing an orthopositronium with a mirror orthopositronium.

orthopositronium has a definite mass. They are mixed together in a very weird way and are no longer separate particles. Such a bizarre situation is already known to happen with strange particles called kaons. Kaons and anti-kaons are short lived particles which were discovered in the 1950's and have since been very well studied in laboratory experiments. The kaon and anti-kaon have the same mass, but there is an effective potential arising from the weak nuclear force which mixes kaons with anti-kaons so that neither a kaon nor anti-kaon has a definite mass.

In quantum mechanics, each particle has an associated wave function. A wave function is a pair of numbers, called a 'complex number', which describe the state of the system. However, for mixed systems – such as orthopositronium or kaons (and maybe neutrinos too...), we need two wave functions to describe them. Moreover, as the system evolves in time, these two wave functions can interfere with each other. The upshot is that a pure orthopositronium state can 'oscillate' into a mirror orthopositronium state. What does this statement mean exactly? The laws of physics are described in mathematical terms which usually allows us to say things in a precise way. The oscillation effect can be calculated and a formula relating the probability that an initial orthopositronium would remain an orthopositronium or be transformed into a mirror orthopositronium can be given[*]. The result is illustrated in **Figure 5.6** on the following page.

Technically, the oscillations do not change the mean lifetime of O at all (let's call orthopositronium simply O and mirror orthopositronium O' to save paper). O and O' decay at exactly the same rate due to the mirror symmetry. However, in practice, the decay of O' is not observed because it decays into mirror photons which escape from the detector without interacting with it. Thus, if O oscillates into O' which subsequently decays into mirror photons, then O *appears* to disappear faster leading to a slightly shorter apparent lifetime. Furthermore, if we keep track of how many positrons we have made, and we count the number of ordinary decays, then we will end up with a discrepancy – since not every positron will lead to an observable burst of energy. These two distinct effects can both be looked for in an experiment.

There is one final important piece of information that I need to mention before I tell you the results of the experiments. When oscillating systems undergo frequent collisions, the oscillations become quite ineffective. Why? This is difficult to explain without giving a course of lectures on quantum mechanics. Very briefly the main points are: An initial O state oscillates into a sort of mixture of O and O'. In a collision the state must behave either as an O or O'. Collisions act a bit like a 'measurement' which fixes the state to

---

[*]Technically, the mathematical formula for the oscillation probability of orthopositronium into its mirror partner is given by, $P(O \rightarrow O') = \sin^2 \frac{\pi t}{t_{osc}}$.

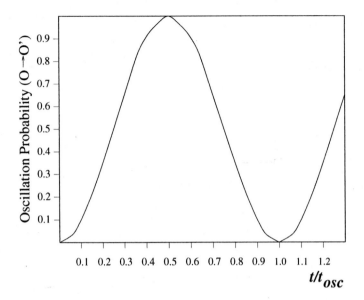

Figure 5.6: The photon-mirror photon transition force causes a strange quantum mechanical effect whereby an initial orthopositronium state (O) has a certain probability of being converted into a mirror orthopositronium state (O'). In fact, this probability actually oscillates over time (t). Because the orthopositronium decays rapidly it turns out that this oscillation effect is small because no orthopositronium particles live long enough to undergo even a full oscillation. That is, t is always much less than t_{osc} so that $P(O \rightarrow O')$ is only slightly greater than zero.

---

be either O or O' but not any mixture. The result is that collisions disturb the oscillations; the oscillations stop and must start all over again after each collision. This means that the oscillations can never have a significant effect if the collisions are frequent enough. This effect has been called 'the watched pot never boils effect', although it is not quite the same effect observed by impatient cooks.

Time to consult with experiments

The orthopositronium lifetime can be measured in a number of different ways. I will talk more about some of these methods in a moment. For now, let me give you the results of the experiments, the most accurate of which are given in the table below[35]

Method	Date	$\tau_{oPs}(exp)/\tau_{oPs}(theory)$	τ_{coll}
Vacuum	1990	0.9988 ± 0.0002	$\sim 0.3\tau_{oPs}$
Gas	1989	0.9984 ± 0.0002	$\sim 10^{-3}\tau_{oPs}$
Powder	1995	1.0000 ± 0.0004	$\sim 10^{-4}\tau_{oPs}$

Table 5.1: The most accurate measurements of the orthopositronium lifetime (third column). The fourth column is the average time between collisions of the orthopositronium with the background particles in each experiment which is an important distinguishing characteristic of these measurements.

In this table I have divided the experimentally measured value of the lifetime with the theoretically calculated value given earlier. Perfect agreement would then correspond to a value of 1.000 for this ratio. Evidently, the powder experiment agrees with the theory prediction while the vacuum cavity and gas experiments do not. These disagreements may not look like much – only about 1/10 of 1 percent. However, for electromagnetism things are routinely measured and calculated to much higher accuracy. The important point is that the small discrepancy has to be compared with an even smaller uncertainty (or error). Generally, if the discrepancy is less than about two or three times the uncertainty there is no reason to get excited about it. However, if the discrepancy is more than four times the uncertainty, as it is for orthopositronium, things begin to get very interesting.

What Mirror Matter can do for you

Let us now examine the possible effects of mirror matter. The first thing to note is that experiments in which the orthopositronium undergoes many collisions will *not* be affected significantly by the oscillations of orthopositronium into mirror orthopositronium. The collisions prevent the oscillations from occurring. Thus, it seems that the experiment which is most sensitive to the mirror matter effect is the vacuum cavity experiment, because only this experiment has a low collision rate (the collision rate is *inversely* proportional to the time between collisions, τ_{coll}, which is given on the previous page in the fourth column of table 5.1). The fact that the measured mean lifetime in that experiment is *roughly* the same as the theoretical value indicates that the mean lifetime must be much less than the oscillation time. This is, of course, possible because the oscillation time is essentially a free parameter; it depends on the strength of the photon-mirror photon mixing parameter (ϵ), which is itself theoretically undetermined [something like the electric charge of an electron (e), which is to be measured in an experiment].

We can conclude that the oscillation effect can *only* be significant for the vacuum cavity experiment where it leads to an effective *decrease* in the orthopositronium lifetime; the precise amount of the decrease depends on the strength of the photon-mirror photon transition force, ϵ. In other words, the existence of mirror matter can explain the decrease in the lifetime for the 'vacuum' experiment (first row of table 5.1), but in doing so it means that the effects of oscillations in the other experiments must be negligible. This is fine for the powder experiment because it agrees with the theoretical result without oscillations. However, for the gas experiment no mirror matter explanation can be given. I reached this conclusion during mid-1998 and decided that the situation was too muddled to publish anything on it. My interest was revived in January 2000 when Sergei Gninenko, from the CERN laboratory in Europe, e-mailed me suggesting that we might collaborate on a joint work examining the experimental situation on orthopositronium and whether or not it supported the mirror matter effect. I promptly replied that I had looked into this effect already but that the gas experiment seemed to be a bit of a stumbling block...Whereupon Sergei, a first rate experimentalist and theorist, told me that he had not only reached a similar conclusion but that a number of problems had arisen with the gas experiment...

Gas problems lead to another mirror matter triumph

Let us take a closer look. In the gas experiment, positrons are produced from the β-decay of a radioactive source. These positrons are given directions by a magnetic field and produce orthopositronium when they are stopped in a gas such as nitrogen. The orthopositronium can either decay or it can collide with a gas molecule after which the positron (inside orthopositronium) can annihilate with an electron (from one of the atoms inside the gas molecule). This latter process, which has nothing to do with the lifetime of isolated orthopositronium, is called the 'pick-off' rate. It has to be somehow estimated in order to determine the true lifetime. [The lifetime values given in Table 5.1 are, of course, the inferred 'true' lifetime of isolated orthopositronium].

In the gas experiments, the idea is very simple. If the presure of the gas is halved, then the number of collisions of the orthopositronium with the gas molecules should also halve which means the pick-off rate will be half what it was. Thus, if we measure the total decay rate at varying pressures we can extrapolate to zero pressure since we know that the pick-off rate is proportional to the pressure (at small pressures....). So far so good. The problem though is that one needs to assume that the orthopositronium has the same temperature as the gas molecules at the lowest pressures used. Initially, the orthopositronium has an independent velocity distribution but after many collisions it 'thermalizes' which means that it has the same temperature as the surrounding gas molecules. However, at the lowest pressures used in the experiment, some studies indicate that the orthopositronium does not undergo a sufficient number of collisions to properly thermalize. This means that the pick-off rate is no longer simply proportional to the pressure. If the pressure is halved, then the pick-off rate is slightly more than half of what it was because the velocity of the orthopositronium will be somewhat higher at low pressures (where it is less thermalized). The effect of a higher pick-off rate at low pressures will tend to make the inferred (isolated) orthopositronium decay rate slightly larger and hence the lifetime slightly shorter than its true value. In other words, there may be nothing wrong with the experimental gas measurements themselves, the problem is that one of the assumptions was not valid, which affected their inferred value of the lifetime. Several groups have suggested that the gas experiment should therefore be ignored.

In contrast, nobody has ever found any problem with the vacuum cavity experiment. In this experiment the pick-off rate is very low since the orthopositronium can only get picked-off when it collides with the walls of the cavity. It is therefore a simpler experiment, which is one of the nice things about a vacuum. The end result is that there is just one *discrepancy* and one *agreement* that needs to be explained. And that is the discrepancy between the cavity experiment and the theoretical prediction, and the agreement between the powder experiment and the theoretical prediction.

If this is the case, then mirror matter does the job so nicely. The large rate of collisions of the orthopositronium in the powder experiment will render oscillations of orthopositronium with its mirror counterpart ineffective. Thus, it should agree with the predicted value from standard physics and it does. The *agreement* is explained. In addition, the vacuum-like conditions in the cavity experiment allow oscillations to have an effect implying a shorter effective lifetime. A result which indeed happens. The *discrepancy* between the vacuum cavity experiment and theory is also explained.

The size of the effect for the vacuum cavity experiment is related to the size of the kinetic mixing parameter, ϵ, which governs the strength of the photon-mirror photon transition force. This parameter is not predicted in the mirror matter theory, but this is not unusual. Every force has at least one free parameter: gravity has Newton's constant, electromagnetism has the electron's charge, and the nuclear forces have free parameters as well. For the photon-mirror photon transition force, the free parameter is ϵ. Anyway, the vacuum cavity experiment provides a type of 'measurement' of this free parameter, which is $\epsilon \approx 10^{-6}$. Of course no-one likes free parameters. That they exist probably suggests that there is some deeper theory out there – perhaps a theory of everything, but what it is nobody knows... It is part of the 'great ocean of truth' which Newton glimpsed on the horizon nearly three centurys ago.

Although the mirror matter theory cannot predict the *size* of the effect (because of the *a priori* unknown parameter ϵ) it does predict the *sign* of the effect. The vacuum experiment is unambiguously predicted to measure a shorter apparent orthopositronium lifetime. If it were observed to have a longer lifetime no mirror matter explanation would be possible.

Anyway, it seems that the orthopositronium lifetime puzzle can be explained if mirror matter exists and is coupled to ordinary matter through the photon-mirror photon transition force. Orthopositronium it seems is an interesting window on the mirror world whose effects can show up in relatively simple laboratory experiments. Indeed, it offers a unique way of really proving that mirror matter exists. However, while the mirror matter theory can nicely explain the orthopositronium lifetime puzzle, this puzzle is based only on one anomalous vacuum cavity experiment. Before one can absolutely believe in the existence of mirror matter on the basis of this experiment one would have to do two things. First, show that there is no other explanation and second, to recheck the experiment again, preferably in several different ways.

If the experiment was repeated with a larger cavity then the mirror matter effect would be larger. In the vacuum cavity experiment the largest cavity size used was such that the orthopositronium (typically) collided with the walls about three times before it decayed. If the cavity was three times larger, then the mirror matter effect would also be three times larger. In other words, instead of finding a lifetime of about a tenth of one percent shorter they should find a lifetime about a third of one percent shorter. Alternatively, if the orthopositronium were moving more slowly, then this would have the same effect as making the cavity larger. Very recently, I have been informed by David Gidley that the University of Michigan group intends to repeat the vacuum cavity experiment more sensitively and in such a way as to search for the predicted mirror world effect.

Another completely different, but complementary way to check things is by using a somewhat different type of experiment designed to search for invisible decay modes. Such an experiment would be really important, since it could actually prove that the orthopositronium decays invisibly. The idea is that the positrons (produced from the β-decay of a radioactive source) are 'tagged' before they reach their target and produce orthopositronium. That is, their presence is registered by means of a counter which keeps track of the time when the positrons are produced. Surrounding the source, counter, and target, is something called a 'calorimeter' which measures the energy deposited when the orthopositronium decays into photons. If such 'tagged' positrons form orthopositronium which subsequently decays invisibly then no energy will be detected. Such 'missing energy' events would then tell us that orthopositronium decayed into something which passed through the detector without producing any energy.

Several such 'tagged positron' experiments have been carried out which appear to exclude invisible decays as an explanation for the faster decay observed in the vacuum cavity experiment. The oscillations into mirror orthopositronium and its subsequent decay into mirror photons is also an invisible decay mode. However, the particular experiments which searched for the effect of missing energy were not themselves vacuum experiments. For this reason they were *not* sensitive to the oscillation mechanism that mirror matter predicts. Nevertheless, the idea of these experiments is very clever and they should detect the missing energy if they were repeated with a vacuum set up. My mirror matter collaborator, Sergei Gninenko (who is, paradoxically, made of ordinary matter!) has been trying hard to secure some funding for this important experiment, but so far without success...

Implications of the photon-mirror photon transition force...

There are several interesting ramifications of the photon-mirror photon transition force. One effect is that it may actually make mirror stars visible if they are able to capture ordinary matter as they traverse through interstellar space. The captured ordinary matter can interact with the hot mirror matter which makes up the mirror star via the small photon-mirror photon transition force. The fundamental process is illustrated in the interaction diagram in Figure 5.2 and similar diagrams where the electrons and mirror electrons are replaced with protons and mirror protons.

In this way, energy can be transferred from the hot mirror matter to the captured ordinary matter which can then radiate ordinary photons which we can detect. The net effect is that mirror stars can potentially be observable if they have a significant amount of ordinary matter within them. One might worry that this would make the interpretation of the 'isolated' planets as ordinary planets orbiting mirror stars (discussed in chapter 4) problematic since this relied on the ordinary stars being invisible. However, these systems were observed in star clusters where they are still very young – only about a million years old. [Recall that one million years is very young for a star especially when compared with the age of the galaxy which is of order ten billion years old]. Mirror stars in star clusters may be too young to have captured much ordinary matter from interstellar space.

Actually, even if mirror stars capture a significant amount of ordinary matter, there are theoretical reasons (to do with the way in which energy is transported through a star) to believe that the effective surface temperature of the small ordinary matter component of a mirror star will be much lower than the effective surface temperature of the mirror matter. This means that mirror stars should be quite dim, and therefore difficult to detect. This could also explain why we don't see the halo mirror stars – if they do indeed exist – as I discussed in chapter 3. It is also possible that mirror stars in the halo might be mostly in the form of dead mirror stars called mirror white dwarfs – which are also very dim, yet potentially observable and may even have been observed!

White dwarfs are very different from standard nuclear fuel burning stars. White dwarfs have extremely hot and dense interiors which do not produce energy from nuclear reactions. They therefore do not have a huge photon flux transporting energy outwards through the object. Because of these differences, the effects of a small ordinary matter component captured by a mirror white dwarf would be quite different to if it were captured by a standard nuclear-fuel burning mirror star. Ordinary matter accreted by a mirror white dwarf would mix with the mirror matter in the 'atmosphere' which surrounds the hot and dense interior. In this case, the effective surface temperature of the ordinary and mirror matter components need not be so different (as they would be for a standard nuclear fuel burning star). Remarkably, mirror matter white dwarfs may have already been observed but interpreted as an unusual type of ordinary white dwarf star...

There have been some interesting, but controversial claims over the last two years or so. Several groups have claimed to have detected cool, faint, white dwarf stars which belong to the halo[36]. Although they are only observed nearby, within 100 light years from us, they can tell that they belong to the halo because of their direction of motion. Their number appears to be consistent with the expectation from the MACHO experiments, and thus may compose up to half the mass of the halo. However, as I discussed in chapter 3, there are significant problems if these objects really are white dwarfs. An interesting alternative possibility is that they are actually mirror white dwarfs. They then have the rough appearance of ordinary white dwarfs simply because we are detecting the light from embedded ordinary matter...

Gamma Ray Bursts and exploding mirror stars

Another possible consequence of the photon-mirror photon transition force is that exploding mirror stars (mirror supernova) should produce an ordinary photon burst as well as a neutrino burst. Recall our earlier discussion about supernova in chapter 3. In that chapter, I mentioned that large stars do not simply fade away when they get old; after they finish their nuclear fuel they collapse with a bang! When the core of an exploding *ordinary* star collapses the temperature becomes so hot that pairs of electrons and positrons are created along with neutrinos and anti-neutrinos.

In the alternative case of a collapsing *mirror* star, pairs of mirror electrons and mirror positrons will be created instead. These particles will produce ordinary electrons and ordinary positrons through the photon-mirror photon transition force (**Figure 5.7**). These electrons and positrons can then get together and annihilate to produce ordinary photons, a process shown in **Figure 5.8**.

The photons, electrons and positrons cannot immediately escape though – they are trapped by their own interactions for a short time. [A similar effect also happens for the neutrinos]. Over a time scale of a few seconds, the photons diffuse outwards, potentially carrying off as much energy as the neutrinos. The net effect is an expanding relativistic fireball leading to a huge photon burst with energies in the MeV range – gamma ray energies[*]. Remarkably such huge photon bursts are indeed observed and are called 'gamma ray bursts'.

Gamma ray bursts were first discovered by US military satellites in 1967. These satellites were designed to monitor Soviet compliance of the 1963 nuclear test ban treaty by looking for sudden bursts of gamma rays of terrestrial origin. They found lots of gamma ray bursts but these were coming from above rather than below. Eventually the military realised that these explosions were not caused by the Soviets, and in 1973 this information was released to the public. The origin of the gamma ray bursts has been a great mystery ever since. It is very tempting to suppose that gamma ray bursts are the observations of exploding mirror stars. This general idea was proposed by Sergei Blinnikov several years ago. More work though is necessary to see if this explanation can explain all of the details of the gamma ray bursts.

---

[*]If a *mirror* supernova emits a burst of ordinary photons then an *ordinary* supernova should release a burst of mirror photons. This is not in conflict with any observations because mirror photons are undetectable to us.

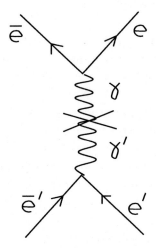

Figure 5.7: The Temperature in the core of a mirror supernova explosion can become so great that mirror electrons and mirror positrons can be produced. Annihilation of mirror electrons and mirror positrons can then produce ordinary electrons and ordinary positrons, via the photon-mirror photon transition force, as shown in the above interaction diagram.

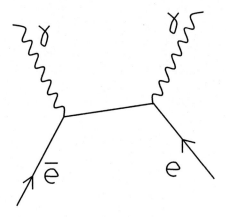

Figure 5.8: Interaction diagram for electron-positron annihilation into two photons.

The good, the bad, and the ugly...

There are many other implications of the photon-mirror photon mixing force, and I will discuss some of them in the next chapter. Clearly though, if the photon-mirror photon mixing force really does exist then it is rather important. Not only does it modify the properties of orthopositronium leading to important laboratory tests, but may lead to interesting astrophysical effects as I have just discussed.

Does the photon-mirror photon transition force suggested by the orthopositronium experiments really exist? Nobody knows for sure. The only way to know for sure is to repeat the orthopositronium experiment in vacuum. This would be the logical thing to do. Unfortunately funding committees are not known for their logic...

Perhaps we should stop to reflect on the way in which science has progressed over the last half-century. An interesting perspective has been given recently by Harry Lipkin in a letter published in Physics Today (July 2000). Entitled *who ordered theorists*, Lipkin writes[37]:

> The science wars that continue in *Physics Today* seem to be only between theorists of various kinds discussing *gedanken* physics. They don't see the real physics as an experimental science that progresses from one experimental breakthrough to another. Theorists are often irrelevant and sometimes actually hinder progress by sitting on committees and opposing the experiments that lead to new breakthroughs. As we begin this new century, "Who needs theorists?" would be an interesting question to ask. How would physics have progressed in the second half of the 20th century – that is, since I received my PhD in 1950 – if theorists had been ignored?
>
> The main breakthroughs in physics since 1950 can be characterised as "who-ordered-that?" effects, named after I. I. Rabi's famous remark about the discovery and existence of the particle we now call the muon...

Lipkin goes on to argue that most discoveries of the second half of the last century were really not anticipated by most theorists. Indeed, the emerging theories were fiercely resisted by the 'theoretical establishment'...

These days in high energy physics, most experiments are very large and very expensive. It is really very unusual that inexpensive small experiments such as the orthopositronium lifetime measurements can shed light on something fundamental like mirror reflection symmetry. It is amusing to recall that the last time something important was learned from a small 'table top' experiment was in 1957 when the Cobalt 60 experiment (mentioned in chapter 1) proved that the interactions of the ordinary particles by themselves appear to violate mirror reflection symmetry. So it would be very amusing if nearly 50 years later another table top experiment showed that mirror reflection symmetry is really a symmetry after all because of the existence of mirror particles. Actually, as I have mentioned such an experiment has already been done in 1990 which does suggest just that, but it clearly needs to be confirmed by independent experiments to make sure. The most important and decisive experiment would be the tagged positron experiment proposed by Gninenko. It would either prove or disprove the existence of invisible orthopositronium decay modes...

Most research in particle physics is directed to trying to prove that supersymmetry and grand unification are correct. As I briefly mentioned in chapter 3, supersymmetry is the peculiar broken symmetry connecting ordinary particles with hypothetical 'superparticles' of different spin. While Grand Unification Theories (or GUTs) are a type of unified theory interconnecting the three non-gravitational forces. (I will talk more about GUTs in Chapter 8). So far, no experiment has ever found any evidence that such things really exist despite huge experimental efforts over nearly a quarter of a century. While grand unification is a good idea, and supersymmetry less so, the lack of any experimental support does suggest that these ideas do not agree with nature. At the very least there is no reason to think that it is likely that they are correct.

Physics is not like mathematics. The aim is to describe nature, not merely to describe possible ways in which nature might be and then to try and force nature to conform to these possibilities. Fortunately though, some of these expensive large experiments can potentially discover other things so that it often happens that they really are very useful. An interesting example of this comes in chapter 7,8 where large and difficult underground experiments were originally constructed to 'verify the existence of grand unification' by trying to find proton decay, but found something else instead.

The problems facing high energy physics have several causes. But I believe that the primary problem is not the lack of government funding but rather a lack of scientific leadership. Many senior scientists behave more like fortune tellers and prophets... They routinely use unscientific arguments and behave as if they have a pipeline to God. In reality though the lines appear to have been crossed and their pipelines seem to come straight from the sewage factory. My colleague Henry Lew and I summed up the situation in 1994 when we wrote (out of sheer frustration with the high energy physics community)[38]:

> The standard model of particle physics (SM) is an extremely successful description of past and existing experiments. At present there is no experimental indication of a deficiency in the SM. In view of this it is not surprising that there is divided opinion about what lies beyond the SM. However, a diversity of opinions on how to connect the known to the unknown is not necessarily a bad thing given that there exists no guaranteed way of making this connection. On the other hand, a reading of the current literature on particle physics could give the mistaken view that the ideas of grand unification and supersymmetry are so well motivated and unique that it is only a matter of time before they are found to be true. Indeed, some people are even advocating spending billions of dollars to test some of the parameter space of some supersymmetric grand unified models[1]*. We feel that this attitude to physics beyond the SM is an extremely biased interpretation and extrapolation of existing knowledge. Unfortunately, this bias in the literature has remained unchecked and has been (and still is) detrimental to the advance of particle physics (and to the tax payer) in terms of both theory and experiment. It is detrimental to theoretical physics in

---

*Ref.[1] is the paper: T. Kamon, J. Lopez, P. McIntyre, J. White, hep-ph/9402349 (go to http://xxx.lanl.gov/format/hep-ph/9402349 to download this article).

the sense that the number and variety of hypotheses generated to guess at the new physics becomes suppressed. The situation for experimental physics is no better because the bias in theory leads to a bias in experiments. Finally, it is tax payer unfriendly because the proponents of supersymmetric grand unified theories demand large accelerators to be built which are currently very expensive.

The current situation is quite different to the era before the discovery of the W and Z gauge bosons. The theoretical case for the existence of the W and Z gauge bosons was very strong. Their masses could be approximately predicted from the data already obtained in low energy experiments. In this case there were strong physics reasons to build the necessary colliders to study these gauge bosons. Unfortunately, the standard model works so well, that there is, at present, no experimental evidence for new physics beyond the standard model. Of course, this does not mean that there is no new physics beyond the standard model. In all likelihood there is new physics, but there is an infinite number of possibilities for what this new physics might be. In light of this current situation, it is rather unimaginative of the particle physics community to spend so much effort repeatedly studying the possibility of supersymmetric grand unified theories. This does little to advance the theory of elementary particles. In our opinion, these theories are uninteresting. Grand unification can be motivated from experiment because they can simplify the gauge quantum numbers of the quarks and leptons but they are uninteresting because they involve new physics at untestable energy scales. Supersymmetry on the other hand is perhaps testable but is uninteresting because it is not well motivated by experiment. *It doesn't explain the existence of any known particle or symmetry.*

In view of the above, we feel that it is important to search for interesting new ideas for new physics beyond the SM (rather than work on the same boring idea over and over again). If particle physics is to advance new ideas are clearly needed. One approach that we have been studying is the issue of the origin of parity violation in nature...

Incidentally I do believe that expensive large colliders should be funded such as the Large Hadron Collider (LHC) at the European Centre for Nuclear Research (CERN). However, the purpose of these machines is not to verify the predictions of supersymmetry or GUTs.

Indeed, these machines will (most likely) *not* find superparticles because there is little evidence that these particles actually exist – however they may find something else (perhaps a who-ordered-that particle?). These are the only machines that can probe physics at higher energies. Nobody knows what they will find and that's the really exciting part....

We all use the benefits of fundamental research. Washing machines, refrigerators, Clint Eastwood films are all made possible because people kept asking why, doing experiments and investigating the basic properties of nature. It is of course impossible to predict the future course of science and technology and one cannot envisage what current knowledge will eventually make possible. However, we can learn from history. Clearly, pioneering workers such as Charles Coulomb and Hans Christian Oersted studying the phenomena of electricity and magnetism, once basic research, couldn't possibly have envisaged its most important application – the washing machine (sorry Clint!).

However, the problem at hand is to find the relatively meager resources necessary to fund the orthopositronium experiment. Only a few hundred thousand US dollars are required. (By comparison the afore mentioned Large Hadron Collider currently being constructed at CERN costs more than 1 billion US dollars....). So, as John F. Kennedy almost said – let us not ask what mirror matter can do for you, but what you can do for mirror matter. If there are any wealthy people willing to help fund the orthopositronium experiment please let me know. We can name a mirror particle after you – especially if your name happens to be 'mirror electron' or 'mirror proton'!

Before I get myself into any more trouble with the establishment, let us continue with the physics... Speaking of which there is one more very explosive implication of photon-mirror photon transitions which the orthopositronium experiments suggest – and that is the subject of the next chapter.

Chapter 6

Mirror Matter and the Tunguska Event

In chapters 3 & 4 I discussed the evidence that the Universe, and our galaxy in particular, are full of mirror matter. Specific evidence for mirror stars and mirror planets was presented. However, I also pointed out that there was not much room for a large amount of mirror matter in the inner part of our solar system. Unseen matter still affects the motion of orbiting bodies such as the planets. Nevertheless, we do not need a large amount of mirror matter in our solar system to have explosive implications for life on our planet. We don't know enough about the formation of the solar system to be able to exclude the possibility of a large number of mirror matter space-bodies if they are small, like comets and asteroids. The total mass of asteroids in the asteroid belt is estimated to be only about 0.05% (that's 1/2000) of the mass of the Earth. A similar number (or even greater number) of mirror matter bodies, perhaps orbiting in a different plane, or even spherically distributed is a fascinating possibility. Even more interesting is that there is evidence that such mirror bodies actually exist and collide with our planet – with devastating consequences.

What the Tungus tribesmen saw

Something very strange happened in Siberia in the early hours of June 30, 1908. Siberia is a vast region stretching from the Ural mountains in the west to the Pacific ocean in the east. Most of the year it is covered by a blanket of snow, but June 30 was a clear day, the snow had melted and the brief summer had begun... In a remote corner of this wilderness, the Tunguska river was winding its way through a peaceful valley. Reindeer were grazing among the pine trees of the Taiga forest when something rather unusual happened – a tremendous explosion occurred...

At the time the world knew nothing of this. Following the Tunguska explosion colourful sunsets and sunrises were the only evidence that something unusual had happened. The *New York Times* of July 3, 1908 reported "remarkable lights" being "observed in the northern heavens on Tuesday and Wednesday nights." It was reported that scientists attributed the dazzling displays to solar outbursts causing electrical disturbances in the atmosphere, but what would scientists know...

Tungus (also called 'Evenki') tribesmen and Russian fur traders knew better. Many of these people had heard, some had seen and a few had felt the effects of the blast. One eyewitness in Vanavara (the nearest village to the centre of the explosion) stated[39]:

> 'The sky split apart and a great fire appeared. It became so hot that one couldn't stand it. There was a deafening explosion and my friend S. Semenov was blown over the ground across a distance of three sazhens (six metres). As the hot wind passed by, the ground and the huts trembled. Sod was shaken loose from our ceilings and glass was splintered out of the window frames.'

Disturbances in the Earth's magnetic field were reported 900 kilometres from the epicentre in Irkutsk. These were "magnetic storms" similar to the ones that occur when a nuclear bomb is exploded in the atmosphere. Tremors from the explosion were recorded in St. Petersburg and in other locations around the world. The vastness of Siberia did not make it easy to investigate what had happened. The last years of the Romanov dynasty were a period of great social up-heaval and disorder, culminating in the Russian revolutions of 1917. It was not the time to investigate mysterious explosions in a remote corner of Siberia.

Exactly what happened on June 30 1908 in Siberia is not known for sure (even today). That something significant happened at all may never have become known to the outside world except for the efforts of one man. His name was Leonid Kulik, and 19 years after the event he decided to try and find out what had happened.

What Leonid Kulik found out

From 1920 Kulik worked at the Mineralogical Museum in St. Petersburg. Part of his duties was to locate and examine meteorites that had fallen within the Soviet Union. Almost by chance Leonid Kulik came across reports that a large fireball had been observed in the sky in the Tunguska river region of Siberia some years earlier. He was curious and suspected that a large meteorite had fallen and wanted to find it. After an initial 'fact finding expedition' in 1921-22 during which he interviewed many eyewitnesses of the fireball, he became more and more convinced that something very unusual had happened in 1908. Among the information he obtained was the following report[40]:

> The Evenki people, who were camping on the Podkamen-naya Tunguska River, describe how strong air pressure laid low a strip of forest, and killed several reindeer belonging to one of the Evenki, Ivan Il'yich, at a camp spot between the Podkamennaya and Nizhnyaya Tunguska Rivers.

From the information gathered, Kulik was able to figure out roughly the location of the 'impact site' which he determined to be just north of a small and remote trading post called Vanavara.

Early in 1927 Kulik and his assistant Gyulikh set out for Vanavara, which is now known to be nearly 100 kilometres from the epicentre. It took nearly two months for Kulik to reach Vanavara. At that time, the Vanavara trading post consisted of only a few houses belonging to the families of several fur traders. From there Kulik interviewed the fur traders and Evenki hunters about what they could recall about the blast of 1908. Many of the locals did not like to talk about the strange happenings in 1908 because it was believed that the events of that year were due to a visitation from their God – Ogdy. They were afraid to enter the area of the blast which was thought to be cursed. Such stories increased Kulik's interest and he was even more determined to continue his expedition. From Kulik's discussions with the natives he managed to determine more precisely the direction of the 'impact site' and began his journey once more, taking with him several guides from Vanavara.

The Tunguska river region is riddled with swamps, bogs, trees and especially mosquitoes. After courageously battling with these mosquitoes for nearly a hundred kilometres, Kulik and his party finally reached the beginnings of the flattened forest. Uprooted trees lay with their tops facing south, that is, the direction from which Kulik had come. In poetic language, Kulik wrote in his diary[41]:

> The north banks of the River Makirta are broken up by knolls like 'little chimneys' that stand out picturesquely against the sky and the Taiga, their almost treeless snow-capped tops stripped bare by the meteorite whirlwind in 1908.

Continuing northwards, Kulik saw a surreal landscape of kilometre upon kilometre of smashed trees – each toppled like matchsticks (see **Figure 6.1**). Later recalling his first impressions he wrote[42]:

> The results of even a cursory examination exceeded all the tales of eyewitnesses and my wildest expectations.

Kulik was keen to reach the centre of the destruction, however his local guides were afraid and refused to continue the journey. With limited supplies and unfamiliar with the area, Kulik thought it too risky to continue alone with his assistant. He went back to Vanavara and some days later returned to the devastated area and travelled on towards the centre of the destruction. As he travelled closer to the epicentre Kulik noticed that the trees had been burned from above. He was sure that this was the result of a sudden flash

Figure 6.1: The felled trees as seen by Kulik during his 1928 Tunguska expedition. (Photograph courtesy of the University of Bologna website: http://www-th.bo.infn.it/tunguska/).

of intense heat. Kulik found the epicentre, the area in that vicinity he called the 'Great Cauldron' which he described[43]:

> I pitched my second land camp and began to circle the mountains around the Great Cauldron; at first I went towards the west, covering several kilometres over the bare hill crests; the tree-tops of the broken trees already lay facing west. I went round the whole cauldron in a great circle to the south, and the broken trees, as though bewitched, turned their tree-tops also to the south. I returned to camp, and again set out over the bare hills to the east, and the broken trees turned its tree-tops in that direction. I summoned all my strength and came out again to the south, almost to the Khushmo: the tops of

the broken trees also turned towards the south. There could
be no doubt. I had circled the centre of the fall!

On this and several subsequent expeditions, Kulik and others mapped
out the fallen trees (see **Figure 6.2** below).

The scale of the destruction was truly enormous – no less than
2,100 square kilometres of destruction. At the epicentre was a large
swamp. On every side of the swamp the fallen trees lay with their
tops pointing outward – like spokes on a giant wheel with the swamp
in the centre. Unable to find the meteorite, he felt certain that it must

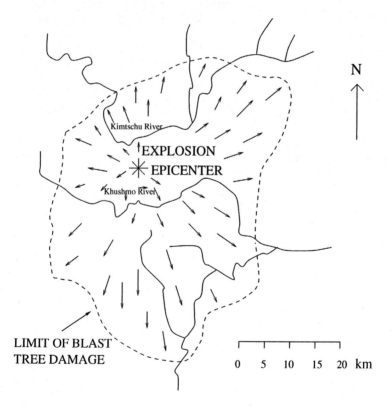

Figure 6.2: The area of fallen trees and their radial orientation at the Tunguska site.

Figure 6.3: When Kulik saw this 'depression' (later named after his assistant Suslov) he became sure that a meteorite fragment fell here. But no meteorite fragment was found in this or any of the other similar depressions. Figure courtesy A. Ol'khovatov, picture taken by V. Romeiko.

reside in this swamp. He also found dozens of 'peculiar flat holes' ranging in size from three metres to about 20 meters in diameter (one such hole is shown in **Figure 6.3**).

He thought that these holes might also contain meteorite fragments. He decided that he would return with digging equipment in an attempt to locate the meteorite and its fragments which he felt sure must be there. Kulik led three more expeditions to the site – in 1928, 1929 and the last one in 1939. However, all of his efforts to find the meteorite or even a single fragment dismally failed. Kulik's work was interrupted by World War Two. He was captured and sadly died of typhus in a Nazi prison camp in 1942.

To date there have been about 40 expeditions to the Tunguska site each unable to find any extraterrestrial fragments[*]. More recent expeditions have used more sophisticated techniques to find cosmic material. For example, a 1999 Italian expedition took cores from the bottom of a lake near the epicentre, only to again find nothing (so far). The size of the hypothetical space-body (as I will discuss in a moment) has been estimated to be between 100 thousand tons to perhaps a million tons. While a large portion of the space-body may have vaporised, one might have thought that some macroscopic fragments would remain. Did this extraterrestrial material simply disappear off the face of the Earth?

Whatever the cause of the Tunguska explosion similar blasts will inevitably reoccur. There is evidence for other but so far smaller 'Tunguska-like' events in the last few years, with one such event occurring in 1997 over Greenland and another occurring this year in Jordan. The origin of these strange events is perhaps one of the most fascinating cosmic conundrums of the 20th century...

The Tunguska space-body

It is widely (but not universally) believed that the Tunguska event was caused by the explosion of a space-body such as an asteroid or comet. When the body reached an altitude of between 2.5 to 9 kilometres there occurred an explosion-like energy release. The TNT equivalent of this explosion has been estimated at around 20 megatons, or about 1000 atomic bombs. There is some evidence indicating that after the explosion at least part of the cosmic body continued its flight on approximately the same trajectory.

---

[*]There have been some recent searches for microscopic particles in tree resin with some success[44]. However, their tiny abundance is hardly consistent with what might have been expected.

The shockwave from the explosion devastated about 2,100 square kilometres of the Taiga forest and the flash burned vegetation over an area of about 200 square kilometres. This was followed by a major forest fire covering an area comparable to that of the devastated forest. The trees were felled radially outwards – like spokes on a wheel, suggesting an approximately spherical shockwave. Investigations in the 1960's identified four smaller epicentres within the larger one. Each of these smaller epicentres were identified by their radial tree-fall pattern and each were presumably caused by individual explosions due perhaps to the breakup of the original space-body into fragments.

There are many puzzling aspects of the event. For example the pattern of the felled forest in the immediate vicinity of the epicentre proves unexpected. First, some trees were felled radially, some were not, and a few even survived the explosion! In fact, near the epicentre it might be expected that the forest fall should be dominated by the vertical component of the shockwave, so the fact that some trees were found to be felled in a radial direction right up to the epicentre suggests a very low altitude explosion. These features and others suggests, according to N. V. Vasilyev (who has spent a lifetime studying the Tunguska event) the following tentative conclusion[45]:

> We may tentatively conclude that along with a great energy release from 5 to 5.8 kilometres above the Earth, there were a number of low-altitude (maybe even right above the surface) explosions that contributed to the total picture of destruction.
>   ....It should be emphasised that though the patchiness of the effects associated with the Tunguska explosion has been noted in the literature more than once, its origin has not been discussed. This seems to be due to serious difficulties of its interpretation in terms of the existing Tunguska cosmic body models.

In my view, the most striking feature of the whole Tunguska business is the remarkable lack of any extraterrestrial fragments. This suggests that the Tunguska event was caused by no ordinary space-body, but what about a mirror matter space-body?

A mirror matter space-body?

Imagine that there was some amount of mirror matter in the solar system when it was formed. Maybe there was not enough to form a planet, but there may have been enough to form asteroid or comet sized bodies.

If such small mirror matter bodies exist and happen to collide with the Earth, what would be the consequences? If the only force connecting mirror matter with ordinary matter is gravity, then the consequences would be minimal. The mirror matter space-body would simply pass through the Earth and nobody would know about it unless the body was so heavy as to gravitationally affect the motion of the Earth. However, if there is a photon-mirror photon transition force as suggested by the orthopositronium experiments, then the mirror nuclei can interact with the ordinary nuclei, as illustrated in **Figure 6.4**.

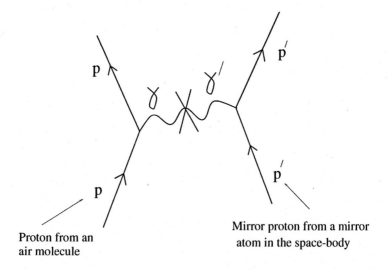

Proton from an
air molecule

Mirror proton from a mirror
atom in the space-body

Figure 6.4: Scattering of air as it passes though the mirror matter space-body. This is the microscopic process responsible for heating the air and the body, and may ultimately be the cause of the explosion of the Tunguska space-body 5 km above the Earth's surface.

In other words, the nuclei of the mirror atoms of the space-body will undergo Rutherford scattering with the nuclei of the atmospheric nitrogen and oxygen atoms. In addition, ionizing interactions can occur (where electrons are removed from the atoms) which can ionize both the mirror atoms of the space-body and also the (ordinary) atmospheric atoms. This would make the mirror matter space-body effectively visible as it plummets to the surface of our planet.

The rate at which the kinetic energy of a space-body composed of mirror matter loses energy through the air depends on a number of factors, including, the strength of the photon-mirror photon transition force (ϵ), the chemical composition of the space-body, its initial velocity and its size and shape. It turns out that for $\epsilon \simeq 10^{-6}$, as suggested by the orthopositronium experiments (chapter 5), the air molecules typically undergo many collisions within the mirror matter space-body. One important consequence of this is that the air resistance (or atmospheric 'drag force') of the mirror matter space-body is actually roughly the same as if it were made of ordinary matter.

We could estimate the initial velocity of the space-body by observing that the velocity of the Earth around the Sun is about 30 km/s. The space-body should have a similar velocity so that depending on its direction, the relative velocity of the space-body when viewed from Earth would be expected to be between about 15 and 70 km/s [*]. In order to account for the huge energy release (which flattened the forest), the Tunguska space-body must be roughly 100 meters in size and weighing of order one million tons. Such a heavy (ordinary or mirror) body would not lose much of its cosmic velocity in the atmosphere if it remained intact. However, if it were to break up into smaller pieces – either via a rapid fragmentation or through some type of explosion – the atmospheric drag force effectively becomes much greater. In that case the kinetic energy of the

---

[*]The minimum velocity of a space-body as viewed from Earth is not zero because of the effect of the local gravity of the Earth. It turns out that the minimum velocity of a space-body is about 11 km/s, for a body in an independent orbit around the sun (and a little less if there happened to be a body in orbit around the Earth).

body can be rapidly dumped into the atmosphere leading to a huge atmospheric explosion.

The resulting picture is that part of the kinetic energy of the mirror matter space-body would be converted into ordinary heat and light as the atmosphere heats up as the mirror space-body passes through it.   Shockwaves would develop as part of the forward momentum of the space-body is transferred to the atmosphere. The body would also heat up *internally* from the interactions of the atmospheric atoms which would penetrate within the space-body. This is quite unlike that of an ordinary body which would heat up only from the friction on the outside. The body would therefore also emit some energy in mirror radiation. More importantly though, the heating up of the whole body could make it easy for the body to fragment and possibly explode. One necessary but surprising feature of the Tunguska event is thereby explained: The atmospheric explosion – necessary in order to convert a large part of the kinetic energy of the body into an approximately spherical down-going shockwave – may arise because the mirror matter body melts and breaks up due to *internal* heating from the interactions of the atmosphere within the body.

Whether this explosion actually happens may depend on many things such as the space-body trajectory, initial velocity and also chemical composition (for example, mirror ice versus mirror iron). The mirror body may have some embedded amount of ordinary matter, so a tiny amount of ordinary extraterrestrial material is possible. Any mirror matter fragments which survive and hit the ground could potentially cause small craters or holes, depending on the size and impact velocity of the fragments. The rate of energy loss in the ground is obviously much greater because the ground is thousands of times more dense than the atmosphere.

Perhaps the most interesting facet of this interpretation of the Tunguska event is that there should be large pieces of mirror matter still lodged in the ground at the Tunguska site, and indeed at other impact sites around the world, as I will discuss. The epicentre itself may have the largest piece... The important point is that the small photon-mirror photon transition force could be large enough to oppose the force of gravity*. Indeed, it might be possible to pick up a piece. I have never given it much thought, but there should be all sorts of applications if such objects could be found – perhaps something even more revolutionary than[46] 'frictionless bearing'! After all, it is a completely different type of matter, with a large possible range of material depending on the chemical composition. For example, mirror H_2O ice, mirror iron, mirror rocks etc., are all possible components of mirror matter space-bodies.

I suspect that mirror matter space-bodies are most likely made from mirror ices, such as mirror H_2O ice. We know that ordinary ices are much more common in our solar system than ordinary non-volatile substances such as rocky materials and metals. Probably this would be true too for mirror matter. An important difference is that mirror ices would *not* be melted by the light from the sun, so that they could in principle be relatively abundant in the inner solar system. If a space-body were made from mirror H_2O ice, any fragments that do not get vaporised during the atmospheric flight would eventually melt after striking the Earth's surface, assuming of course that the air temperature is greater than zero degrees Celsius. Transfer of heat from the air to the body is actually quite efficient. This could explain why no substantial mirror matter fragments were found at Tunguska; most of the space-body had vaporised after it exploded in the atmosphere, any remaining fragments had melted by

---

*Technically, there are two quite distinct cases, depending on the sign of ϵ. (The orthopositronium experiments do not provide any information on the sign of ϵ, for definiteness we have assumed that it is positive, but it may well be negative). Either the photon-mirror photon mixing induces a small ordinary electric charge for the mirror electrons (ϵe) of the same sign as the ordinary electrons, or the sign is opposite. In the first case, the photon-mirror photon mixing force leads to electrostatic *repulsion* between the mirror atoms (of the mirror matter fragment) and the ordinary atoms in the earth. In the alternative case of negative ϵ, there is actually electrostatic *attraction* between ordinary and mirror atoms. In both cases though the interactions should be strong enough to stop a piece of mirror matter from falling through the Earth. However an important difference between the two cases is that in the first case ($\epsilon > 0$) the repulsion will cause mirror matter fragments to remain on or near the surface, largely unmixed with ordinary matter, while in the second case ($\epsilon < 0$) the mirror matter will penetrate the earth (a few metres perhaps) becoming completely mixed in with ordinary matter, (and releasing energy in the process).

the time Kulik arrived there... Once in the liquid state, mirror matter should seep into the ground, probably making its extraction impossible. Still, there should be some proportion of non-volatile material (mirror rock and iron fragments) in the mirror matter space-body. One difficulty is that it might have become mixed with ordinary matter after falling to the ground making its identification more difficult. Still, it should be there...

Whatever the cause of the Tunguska event it is important to realise that there will be other similar events. This is not merely an academic issue: A Tunguska sized explosion occurring randomly over the surface of the Earth could be expected to cause at least 10,000 fatalities on the average. This is because the mean population density over the entire world's surface is currently about 10 people per square kilometre, giving on average, about 20,000 people within the area of Tunguska's devastation. Of course, the number of fatalities could greatly exceed 20,000 people if a Tunguska-sized explosion occurred near a large city.

Clearly, an important issue is to determine exactly how often such huge explosions occur. It is possible that they could occur as often as every 30 years or as infrequently as once every 1000 years. If Kulik had not searched for Tunguska that event may never have been discovered. One lesson is that there could really have been several such events in the 20th century but we just didn't find out about them because they might have occurred over an unpopulated area. Currently more than 85% of the surface of world is unpopulated. Most of the Earth's surface is covered by oceans (71%), some of it is covered by ice (Antarctica, Greenland, Arctic) and there are large deserts and jungles and I haven't even mentioned Siberia.

It is true though, that recorded history in Western Europe, India and Eastern China goes back several thousand years. If such a large explosion occurred in these areas during the last few thousand years we might be aware of it today. However, a Tunguska event occurring on our planet at the average rate of once every 30 years corresponds to a Tunguska event in Western Europe only once every 6,000 years because the area of Western Europe is only about 1/200 of the world's surface area. A similar rate would occur in India and eastern China which are each roughly the same area as Western Europe. Because Tunguska type explosions don't seem to leave any significant craters we just wouldn't be aware of their frequent occurrence throughout the history of the world, if indeed they do occur frequently. Leonid Kulik deserves great credit for his tenacity for 'discovering' Tunguska. I believe that it may have been one of the most astonishing and important discoveries of the 20th century.

If the Tunguska event was caused by a mirror matter space-body, then there should be smaller such events occurring very frequently. In fact there are! Some of these events have been studied in considerable detail by Zdenek Ceplecha and co-worders[47]. These remarkable events have also been catalogued by Andrei Ol'khovatov (www.geocities.com/olkhov/tunguska.htm). Ol'khovatov suggests that these strange events are produced by some poorly understood coupling between tectonic and atmospheric processes rather than by cosmic collisions. However, Andrei didn't know about mirror matter... One of the most recent of these small Tunguska-like events was one that occurred over Jordan.

Mini-Tunguskas: The 2001 Jordan event

On the 18th of April 2001 a fireball was sighted in the sky over Jordan. It was observed to break up into 2 pieces and hit the ground by about 100 local residents in a village about 55 kilometres from the capital Amman. According to the official Jordan Astronomical Society (JAS) report[48]:

> On Wednesday 18 April around 7pm, which is before sunset (sunset occurs around 7:10pm) at sundown to bury a village resident, more than 100 persons saw a bright object moving in the sky with a dark yellowish colour. The object was moving from west to east, and then it broke up into two parts, which fell on a nearby hill (which is about 1.5 km from the place at which we were watching!). As the two pieces hit the ground we saw a fire, initially with a greenish colour, and then the fire reached up to 5 metres! On the very next day I (Mr. Miqdadi) went to that place and I saw the two locations at which the two parts fell. (Let's call the first location A, and the second one B).

Figure 6.5a: A Picture of the Jordan site(s). Courtesy of the Jordan Astronomical Society website (http://www.jas.org.jo/mett.html).

Now JAS is watching and examining the location A, which no one entered yet! The ground is full of ash and it is rather black (from the fire) and so are the stones! What directly brought our attention were two things, the first was a tree trunk which is broken into two parts (See Photo). Mr. Miqdadi said this must be from the object which hit the tree! Actually the appearance of the broken tree trunk is very strange! I don't guess it is a man-made break! The other thing was a half burnt tree (see photo)!

Concerning the location B, which was visited by two persons before JAS, it was also full of ash and black. "The location was full of small rock, but when the object hit the area it made a crater as you can see", Mr. Miqdadai said. Actually there was no real crater! But it was clear that at certain place the level of the rocks is lower than the surrounding, and there is a shape of an arc. Also, a half of a large rock was burnt, while the other half is normal (See Photo)!

We did our best to find a meteorite but I must say that we failed! So the question is what fell then? Did the object totally burn up? Is this ash the meteorite remnant! Eng. Khalil Konsul said, this is not possible, because if the ash is a meteorite remnant, then the meteorite would be very large and this will make a real trouble! Which was not the case! JAS took a sample of the ash and soil, and we shall send it for analysis.

Figure 6.5b,c: A Pictures of the Jordan site(s). Courtesy of the Jordan Astronomical Society website (http://www.jas.org.jo/mett.html).

Some of the pictures are shown in **Figure 6.5**. For more details and for the other pictures, see www.jas.org.jo/mett.html.

Events such as the Jordan event cannot be explained in terms of a space-body made of ordinary matter. The problem is that any large fireball which survives at low altitudes must have originated from a much larger fireball at high altitudes. Impacting space-bodies undergo enormous pressures at their surface when they enter the atmosphere with cosmic velocity and consequently undergo mass loss as the surface melts and vaporises. This effect is called 'ablation' in the technical literature. This extreme heating of the surface is also believed to be responsible for much of the light emitted from (ordinary matter) space-bodies. The Jordan space-body, if made of ordinary matter, should have illuminated a large part of the Middle East which was (inexplicably) not observed to occur.

On the other hand, if the body was primarily made of mirror matter, the rate of ablation of the body would be completely different to that of an ordinary matter body. The pressure of the atmosphere is dissipated some distance within the body rather than just at its surface. Furthermore, in the case of a mirror matter body impacting with the atmosphere, the origin of the light is very different to that of an ordinary matter body. The ordinary light from a mirror matter body comes from several sources, including:

- Interactions with the air through ionizing collisions (where electrons are removed from the atoms).

- The potential build up of ordinary electric charge as a consequence of these ionizing collisions which can trap ionized air molecules within the body. This build up of charge can lead to electrical discharges[49].

- Heating of any ordinary matter fragments within the mirror matter space-body, which subsequently radiates ordinary light.

Even though it doesn't seem to be possible for the Jordan space-body to be made of ordinary matter, we obviously cannot be certain that it was something made of mirror matter. Maybe it was due to some type of geophysical or atmospheric process that somehow took the appearance of a fireball, for example, some type of 'ball lightning' perhaps, or some even more exotic new effect as suggested by Andrei Ol'khovatov and others. Clearly, the only way to be certain that the event was due to a mirror matter space-body would be to dig the mirror matter remnants out of the ground. As far as I know, nobody has ever tried to do this, so it would be worth giving it a go.

Of course, it may not be so easy to recognise a mirror matter fragment as it might become mixed-in with ordinary matter upon striking the ground. What we need is a simple way of extracting mirror matter from the earth. Perhaps one possibility would be to dissolve the sample in acid or alkali, depending on the composition of the material (for example, limestone versus sand), and filter it though a fine sieve. In this way, mirror matter could be left behind... Alternatively we could take samples of organic matter (such as tree parts) and incinerate them at high temperatures, again leaving behind a more concentrated sample containing mirror matter. Even though mirror matter is chemically inert it still has weight, so we can infer its presence provided that the sample is concentrated enough. Let me add that I am no chemistry or materials science expert, so there are likely to be much better ways of purifying and extracting mirror matter from a given sample. Perhaps someone who reads this book may be able to help...

On a cautionary note, let me just mention that it is possible that mirror matter could be hazardous to health. Understanding its properties microscopically is one thing, understanding what it might do to living organisms (such as us) is another. In case anyone is tempted to carry out such an analysis, I would definitely suggest basic safety precautions such as protective clothing to avoid contacting it with your skin and perhaps some type of mask to avoid inhaling it. Of course it may be completely safe (depending on its composition, for instance mirror iron particles versus mirror ice). In any event, I would not like to be responsible for any cases of mirror matter poisoning!

Maxi-Tunguska: The origin of the Moon?

Our Moon occupies an exceptional place among the satellites of all the planets. The reason is because of its large size. The only other terrestrial planet which has moon(s) is the planet Mars. Mars has two moons, called 'Phobos' and 'Deimos'. However, both of these moons are a few hundred times smaller in diameter than our Moon. While it is true that Jupiter and Saturn have moons of comparable size to our Moon – Jupiter and Saturn themselves are much larger planets.

There are a number of other things that we know about the Moon. First is its high integrity – it always shows us the same face. The Moon rotates about its axis in exactly the same time it takes for it to orbit around the Earth. Why? The answer was first figured out by Isaac Newton some time ago... The mutual gravitational influence of the Earth and Moon causes bulges and tides in the Earth and also bulges in the Moon (but no tides of course!). These bulges have the effect of reducing the rotational energy so that today the Moon has stopped rotating completely relative to the Earth. Eventually the Earth will stop rotating relative to the Moon, it's currently slowing down at about 0.001 seconds every century. This rotational energy gets converted into an acceleration of the Moon, which means that the Moon is slowly moving further and further away from us – at about 3 to 4 cm per year. This effect has been measured by bouncing laser beams off the mirrors left on the Moon by the Apollo astronauts. By measuring precisely the time it takes for the light to travel to the Moon and back we can figure out very accurately exactly how far away the Moon is.

Anyway, we can conclude that when the Earth and the Moon were first formed, which is believed to be several billion years ago, they must have been very close to one another. In fact, the material of the Earth and the Moon may have originally formed from a single body which then broke up into two pieces – but how?

A popular theory for the origin of the Moon is that it was formed when a very large asteroid or small planet impacted with the Earth during the early stages of the Earth's formation. One of the problems with this idea is that the chemical composition of the Moon is a bit too similar to the Earth's mantle. There should be a significant amount of extra-terrestrial material left over in the Moon making the chemical composition of the Moon more different to that of the Earth's mantle than it is known to be. However, if the colliding space-body was made of mirror matter than this would alleviate this problem. First, a smaller body may be needed if it was made of mirror matter, especially if the body penetrated deeper into the Earth before releasing its kinetic energy, thereby making it easier to liberate enough material to form the Moon. Second, any mirror material left in the Moon would eventually diffuse toward the Moon's centre before the moon solidifies. In any case, it would be undetectable and the composition of the Moon would then appear similar to that of the Earth's mantle.

*   *   *

If Tunguska and many other explosions are caused by the collision of space-bodies made of mirror matter with the Earth, then this has other important implications. Efforts to protect the Earth against the threat of impact from ordinary asteroids and comets may well be insufficient. Space objects made of mirror matter may potentially pose an overall greater risk than space-bodies composed of ordinary matter. An approaching space-body made of (pure) mirror matter would not be detectable; only after it impacts with the atmosphere would its effects be observable, but then it would surely be too late to do anything about it. Perhaps this is the reason why the Tunguska space-body was not observed in the sky prior to its collision with the Earth. It is also possible that (predominately) mirror matter space-bodies may contain some embedded ordinary matter, whether or not it is enough for the space-body to be observable on its approach to Earth may be an important issue if we want to try to prevent potentially dangerous collisions. Hopefully we will not follow the fate of the dinosaurs which brings us to our next topic.

The extinction of Nemesis?

The evolution of biological species is not always a slow continuous process. There is evidence that there are relatively abrupt points in time where a large number of species are wiped out. The largest extinction occurred about 250 million years ago in which about 80% − 90% of plant and animal species were destroyed. The most famous of the mass extinctions is the one which happened about 65 million years ago. This mass extinction claimed the dinosaurs and about 50% of other plant and animal species.

What is the cause (or causes) of these mass extinctions? This is a very difficult question to answer. The problem with mass extinctions is that they occurred a long time ago. Trying to figure out what caused them is not so easy; all of the eyewitnesses were killed so all evidence is circumstantial. In 1983 a paper[50] by two paleontologists, named Raup and Sepkoski, found something very strange. Their analysis of the fossil record suggested that mass extinctions occur *every* 26 million years − like clockwork. **Figure 6.6** shows the analysis of Raup and Sepkoski. Many people dispute this surprising claim and even today it is not completely clear whether it is really correct.

On another front there is also evidence that the particular mass extinction which wiped out the dinosaurs 65 million years ago was caused by the collision of a large asteroid or comet with the Earth. The evidence is in the form of an excess of the rare element iridium in clay samples dating from that time period. Iridium is very rare in the Earth's crust and mantle but much more common in asteroids and comets. There is also evidence for a large meteor crater also dating from the same time period. It is located in the Yucatan peninsula of Mexico. The estimated size of this asteroid or comet is 10 kilometres in diameter with a mass of about 500 billion tons − about a million times heavier than the Tunguska object! However, at the other mass extinctions there is no observed iridium anomaly or clearly identified craters. If one believes that extinctions are periodic and that they are caused by cosmic collisions, then how can these two things be reconciled? To understand this we need to know something more about comets.

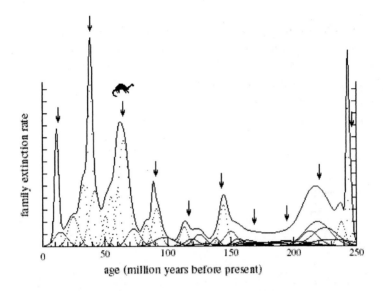

Figure 6.6: The strange results of Raup and Sepkoski which suggested that mass extinctions were somehow periodic, occurring every 26 million years.

Comets can be divided into two classes. Those which orbit the Sun in relatively short periods typically less than 200 years. The most well known such comet is Halley's comet which orbits the Sun once every 76 years. The other class of comets are the 'long-period comets' which have orbital periods greater than 200 years. In fact, most of these long-period comets have orbital periods greatly in excess of 200 years, with values extending up to 10 million years. Of course, we can only see such comets when they enter the inner solar system which means that they must have very high eccentricity. That is their orbits are very elongated ellipses with their closest distance (perihelion) much less than their furtherest distance (aphelion). This is illustrated in **Figure 6.7** on the following page.

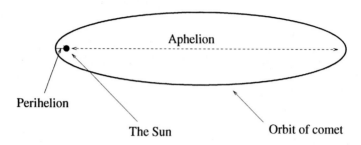

Figure 6.7: The orbits of comets are generally highly elongated ellipses.

The aphelia of the long-period comets typically lie far beyond the orbits of the planets and lie in all directions from the Sun. Long-period comets are believed to come from a comet reservoir called the 'Oort cloud' about $3,000 - 100,000$ AU from the Sun. (Recall that 1 AU is the distance of the Earth from the Sun). By comparison the nearest star is about 280,000 AU from the Sun. It is estimated that there are about $10^{12} - 10^{13}$ comets in the Oort cloud with a total mass of roughly 100 times the mass of the Earth. The Oort cloud cannot be observed directly because the comets are too small and too far away to be observed with even the most powerful telescopes. Nevertheless, we suspect that the Oort cloud really ought to be there because of the properties of long-period comets. Occasionally passing stars perturb this cloud of comets causing some of them to be deflected into the inner solar system. If they enter the inner solar system they can potentially impact on Earth. This can lead to cosmic collisions, but why should they be periodic occurring once every 26 million years?

Imagine that our sun has a dark companion in orbit around it. If this companion were in a very distant elliptical orbit then it could periodically disturb the Oort cloud. When it does this, it could send showers of Oort cloud comets into the inner solar system causing mayhem and destruction. This is the idea behind the Nemesis hypothesis proposed by two groups[51] of US scientists in 1984. However the period (26 million years) is rather long. It means that Nemesis would have an orbit which takes it further than two light years from the Sun (1 light year is 70,000 AU). Whether such an orbit could have lasted for billions of years without being disrupted by passing stars and molecular clouds seems unlikely. Searches for Nemesis have failed to find it. Today most people believe that it probably doesn't exist – and they are probably right. However, it might be possible that it may exist and that Nemesis is a mirror star which might explain why we don't see it. Even if Nemesis is invisible, it might one day be 'discovered' from its gravitational effects (like the MACHO events discussed in chapter 3).

Whether or not Nemesis exists there is an interesting puzzle. Why does the iridium excess appear at only one of the mass extinctions? Perhaps this could be simply explained if the other mass extinctions were caused by the collision of the Earth with mirror matter space-bodies rather than a space-body made of ordinary matter. Such mirror matter objects might also inhabit the Oort cloud. This is a possibility because we have very little knowledge of the composition of the Oort cloud.

Let me now wrap up this chapter with a brief look at a few other speculations which might potentially motivate mirror matter objects in our solar system.

Planet X

Over the years there has been some speculation that their might exist a tenth planet – planet X. As I discussed in chapter 3, Neptune's existence was correctly inferred because its gravitational influence left a tell-tale signature on the orbit of Uranus.

Around the end of the 19th century, residual anomalies in *both* Uranus and Neptune led some astronomers to suspect the existence of a ninth planet with an orbit beyond Neptune. Indeed a dim object was eventually found in 1930 after a long search. It was called Pluto in part because the first two letters are the initials of Percival Lowell. Lowell had initiated the search for the ninth planet a quarter of a century earlier – but died before its discovery. Did Pluto really solve the anomalies in the orbits of Uranus and Neptune? This would only be possible if its mass were big enough – about as massive as the planet Mars or larger. As a sort of historical check, I borrowed my parents' encyclopaedia (1968 Britannica) to see exactly what it says about Pluto in 1968. It states[52]:

> ....Its mass as derived from its attraction on Neptune is about that of the earth, but this value is also uncertain. If Pluto is as large as the earth, as indicated by its calculated mass, its reflectivity must be extremely low. It therefore seems probable that Pluto's diameter is about 0.45 and its mass 0.1 that of the Earth.

To directly determine Pluto's mass is not so easy. If we had something orbiting around it, then its mass could be determined from Newton's Laws. Observations in 1978 revealed that Pluto did indeed have a satellite, which was called Charon. From the orbit of Charon the mass of Pluto could finally be reliably estimated. The result was that Pluto is a planetary featherweight, weighing-in at only 1/500 that of Earth. This is 50 times lighter than the 'probable' value given in 1968, and its diameter is now known to be only about 0.18 that of the Earth (c.f. 0.45 in 1968!). The end result is that the mass of Pluto is several hundred times too small to have any observable effect on the orbits of Neptune and Uranus! Naturally this led to the suggestion that another, heavier planet might exist which could be responsible for the apparent anomalies in Neptune and Uranus.

In 1978 two US astronomers, Robert Harrington and Thomas Van Fladern[53], not only suggested that planet X might explain these anomalies but also argued that planet X may explain the odd behaviour of some of Neptune's moons. Triton – the largest of Neptune's moons – orbits backwards (called 'retrograde' orbit). Another moon of Neptune, called Nereid, revolves in the conventional direction but in an extremely elliptical orbit, ranging between 1.4 to 9.7 million kilometres from Neptune – making it more elliptical than any other moon in the solar system. Harrington and Van Fladern's idea was that maybe Planet X was itself in a highly elliptical orbit which at some point in the past had a close encounter with Neptune. So close, that it reversed the motion of Triton, elongated the orbit of Nereid and knocked Pluto, which Harrington and Van Fladern assumed was originally a moon of Neptune, away from Neptune altogether...

By 1993 the origin of the anomalies in Uranus was clarified to some extent. Voyager 2 measured precisely the gravitational effects of Neptune during its visit to that planet in 1989. It turned out that the mass of Neptune had been over-estimated by 0.5% – that is, about the mass of Mars. With the new value for Neptune's mass, the anomalies in the orbit of Uranus were significantly reduced.

However, there still exists some anomalies in Neptune's orbit. Things are somewhat tricky though. Neptune has not yet completed a full orbit since its discovery in 1846. Interestingly, there are observations of Neptune before 1846 – Neptune had been observed, but people mistook it for a background star rather than a planet. At this point astronomy begins to look a bit like archaeology. The important point is that any anomaly, if real, will become larger over time. Thus, knowledge of precisely where Neptune was, in say 1600, could allow a sensitive test for additional sources of gravity beyond the Sun and the known planets. In fact, there is evidence that Galileo himself observed Neptune in January 1613. This was near the time when Neptune passed behind Jupiter when viewed from Earth. Galileo's positioning of Neptune has been interpreted by some people as evidence for planet X – it has also been interpreted by others as a mistaken observation. Neptune was observed again in 1795 by Joseph-Jerome Lalande a French astronomer, also slightly away from where it should have been – but again it might be some mistaken observation...

We can say that after hundred's of years of observations the evidence for a tenth planet from anomalies in the orbits of the other planets is interesting but still inconclusive. This does not mean that additional planets do not exist. In fact, there is some more recent and independent evidence for a heavy planet or small star in a much more distant orbit where it would not directly affect the motion of Neptune at all.

Goldilocks and some speculations about Pluto, Triton and the Comets

So far, the history of a dark companion to the Sun is somewhat less than glorious. At the moment the evidence for a very distant companion (Nemesis) from the alleged periodic nature of mass extinctions seems very controversial and problematic, while evidence for a nearby planet from anomalies in Uranus and Neptune also appears unconvincing. At this point things begin to resemble Goldilocks and the three bears. The first bowl was too hot, the second one was too cold, but perhaps the third bowl was just right (or perhaps not, maybe Goldilocks should have made her escape when she had the chance....).

The third bowl could be Murray's planet. In the case of Nemesis the Oort cloud is presumed to be perturbed every 26 million years. The good news is that the next mass extinction won't occur until Nemesis passes through the Oort cloud again, which won't happen for at least another ten million years. The bad news is that because Nemesis is now outside the Oort cloud there are no perturbations at the moment, so we don't know if this idea is correct. Ten million years might be a little too long to wait to test the theory.

It is possible that there could exist a distinct companion to the Sun which is *currently* in the Oort cloud, and nothing to do with Nemesis. If there is a large planetary or star sized object within the Oort cloud at the moment then its presence can be revealed from the observed properties of the long-period comets. The idea is that a hypothetical companion within the Oort cloud would deflect small comets from the Oort cloud into the inner solar system. Assuming for simplicity that the companion orbits in a circle, then these comets would originate from a point on this circle. In other words, the signature of a companion star or planet in the Oort cloud would be an observed enhancement or excess of comets coming from the direction of this 'great circle'.

Several people have studied the orbits of long-period comets and in 1999, have indeed found interesting evidence that there is a definite excess of comets on a great circle. The situation is somewhat tricky though. There are two independent studies, one by J. Murray[54] and one by J. Matese, P. Whitman and D. Whitmire[55]. Murray selects a sample of the most accurately observed comets while the second group takes a larger sample. The two studies do not quite agree about the location of the great circle, which is very embarrassing. According to Murray, this companion would orbit at about 32,000 AU from the Sun with an orbital period of about 6 million years. Perhaps short enough to avoid the problems with stability that Nemesis faces. Things would become much more interesting if some independent signature were to be found, for example by gravitational lensing. Until this time, it remains an interesting but as yet unproven candidate for a mirror matter companion to the Sun.

* * *

In examining the case for a mirror planet/star in our solar system we have looked at the evidence for several hypothetical objects. There is actually one other possibility that I have not yet discussed. Paradoxically, the most obvious things are sometimes not so easy to recognise. In fact, it was only while writing this book that the following crazy idea occurred to me. Could any of the known planets or moons be made predominately of mirror matter? As far as the planets go, Pluto is the obvious black sheep of the solar system. Why? Because it orbits in a different plane and has an orbit which is quite elliptical. At its closest approach to the Sun (Perihilion) it is 29.7 AU, while at its most distant point (aphelion) it is 49.3 AU. There are also moons with very strange properties such as Triton, which has a density similar to Pluto and, as previously mentioned in connection with planet - X, orbits Neptune backwards. Perhaps Pluto and Triton are really mirror worlds which accreted ordinary matter on their surface giving them the appearance of an ordinary matter body?

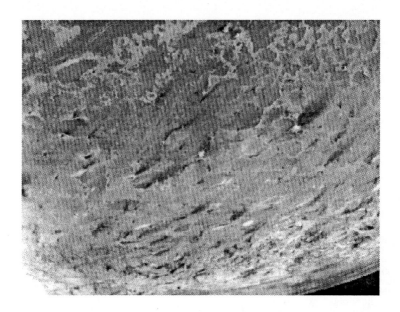

Figure 6.8: Neptune's strange moon Triton. Could Triton be a mirror world? (Credit: NASA, Voyager Project, Copyright Calvin J. Hamilton).

If Pluto and Triton are mirror worlds then their temperature should be less than expected because the surface should thermally radiate both ordinary and mirror photons (that is, ordinary and mirror 'heat'). Since mirror photons are not detected, this should lead to an apparent energy imbalance. In other words, Pluto and Triton should appear to radiate less energy than they absorb from the Sun (and generate from other sources), if they are indeed mirror worlds. NASA's Voyager 2 passed within 40,000 km of Triton in 1989 obtaining interesting pictures (such as **Figure 6.8** above) and temperature measurements. Interestingly Triton's measured surface temperature of $-235\,^oC$ makes it the coldest body yet measured in the solar system. In the case of Pluto, its surface temperature will be studied in detail by the Pluto-Kuiper Express spacecraft mission. That mission is currently planned for launch in 2004 and flying passed Pluto around 2010.

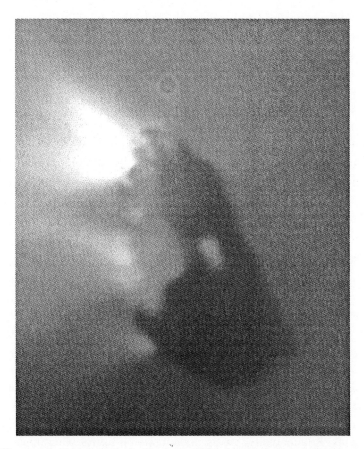

Figure 6.9: Comet Halley's Nucleus. This picture was taken by the spacecraft Giotto. Contrary to prior expectations, Halley's nucleus is very dark, its albedo is only 0.03 making it one of the darkest objects in the solar system – darker even than coal! (Credit: Halley Multicolour Camera Team, Giotto, ESA).

Perhaps many comets and even some asteroids could also be predominately made of mirror matter with some small embedded ordinary matter component. Maybe even Halley's comet is a candidate for such a mirror matter object (**Figure 6.9**)? In fact, I read recently[56] that new comets unexpectedly lose a factor of 100-1000 in average brightness after approaching the sun for the first time, while old comets show little evidence for further significant brightness decreases. Perhaps this could be explained by the heating and subsequent escape of volatile ordinary matter components such as water ice. Remember that only the ordinary matter component is responsible for the ordinary light emitted by the object, and if this turns to gas and escapes then the object would obviously get a lot dimmer...

Of course, it is also possible that all these ideas may be the result of an over-active imagination, or even due to a rare medical affliction in which I 'see' mirror matter everywhere...

What does it mean for you

The fascinating possibility remains that mirror matter bodies exist in our solar system. The explosive evidence from Tunguska and other such events does seem to support the fantastic idea that small mirror matter objects are randomly colliding with the Earth as we go on our seemingly endless journey around the Sun. This interpretation of Tunguska does rely on the photon-mirror photon transition force. This fundamental interaction provides the mechanism causing the mirror body to release its kinetic energy in the atmosphere thereby making its effects 'observable'. Thus, perhaps one way to test the Tunguska hypothesis is to repeat the orthopositronium experiments. If there really is a significant photon-mirror photon mixing force (that is, $\epsilon > 10^{-9}$) then this must show up if careful and sensitive experiments on orthopositronium are done. Alternatively we could go to the impact sites and start digging. Fragments of mirror matter should still be there in the ground if these strange 'meteor events' are in fact due to mirror space-bodies as I have suggested. In cases such as the Jordan explosion, this should be quite easy since the impact area is very well defined and also very accessible. Actually this may be rather important since mirror matter may have all sorts of useful applications...

A distinct possibility is that large mirror object(s) exist with planetary or stellar mass in a distant orbit. It seems to me though, that the evidence for such a large body is not as strong as the evidence for small mirror space-bodies from Tunguska-like events. Perhaps the most interesting evidence for such a large body comes from the observations of Murray and others that many long-period comets appear to originate from a great circle in the heavens. It is most interesting not because the evidence is necessarily stronger, but because it can be further tested in the near future by studying more comets. If the evidence strengthens, then it may be possible to infer the current location of the perturber (our putative mirror star). Its existence could then be tested by gravitational effects such as the amplification of light from stars behind this object. There is also the fascinating but very speculative possibility that Pluto, Triton and some of the comets/asteroids are mirror matter bodies...

One final indication for mirror matter in our solar system comes from the Pioneer 10 and 11 spacecraft anomalies. These spacecraft, which are identical in design, were launched in the early 1970's with Pioneer 10 going to Jupiter and Pioneer 11 going to Saturn. After these planetary rendezvous, the two spacecraft followed orbits to opposite ends of the solar system with roughly the same speed, which is now about 12 km/s. The trajectories of these spacecraft were carefully monitored by a team of scientists from the Jet Propulsion Laboratory and other institutions[57]. The dominant force on the spacecraft is, of course, the gravitational force, but there is also another much smaller force coming from the solar radiation pressure – that is, a force arising from the light striking the surface of the spacecraft. However, the radiation pressure decreases quickly with distance from the sun, and for distances greater than 20 AU it is low enough to allow for a sensitive test for anomalous forces in the solar system. The Pioneer 11 radio system failed in 1990 when it was about 30 AU away from the Sun, while Pioneer 10 is in better shape and is about 70 AU away from the Sun (and still transmitting!).

The Pioneer 10/11 spacecrafts are very sensitive probes of mirror gas and dust in our solar system if the photon-mirror photon transition force exists as suggested by the orthopositronium experiments. Collisions of the spacecraft with mirror particles will lead to a drag force which will slow the spacecraft down. This situation of an ordinary matter body (the spacecraft) propagating though a gas of mirror particles is a sort of 'mirror image' of a mirror matter space-body propagating through the atmosphere which I considered at the beginning of this chapter.

Interestingly, careful and detailed studies[57] of the motion of Pioneer 10 and 11 have revealed that the accelerations of *both* spacecrafts are anomalous and directed roughly towards the Sun, with magnitude, $a_p = (8.7 \pm 1.3) \times 10^{-8}$ cm/s^2 . In other words, the spacecrafts are inexplicably slowing down! Many explanations have been proposed, but all have been found wanting so far. For example, ordinary gas and dust cannot explain it because there are rather stringent constraints on the density of ordinary matter in our solar system coming from its interactions with the sun's light. However, the constraints on mirror matter in our solar system are much weaker because of its invisibility as far as its interactions with ordinary light is concerned.

If this anomalous acceleration of the spacecraft is due to remnant mirror matter gas or dust in our solar system, then calculations of Ray Volkas and myself suggest a density of mirror matter in our solar system of about $\approx 4 \times 10^{-19}$ g/cm^3. It corresponds to about 200,000 mirror hydrogen atoms (or equivalent) per cubic centimetre. If the mirror gas/dust is spherically distributed with a radius of order 100 AU, then the total mass of mirror matter would be about that of a small planet ($\approx 10^{-6} M_{sun}$) with only about $10^{-8} M_{sun}$ within the orbit of Uranus, which is about two orders of magnitude within present limits. If the configuration is disk-like rather than spherical, then the total mass of mirror matter would obviously be even less. The requirement that the mirror gas/dust be denser than its ordinary counterpart at these distances could be due to the ordinary material having been expelled by solar pressure...

In part II of this book the evidence for mirror matter in the heavens was discussed and summarized in Figure 4.13. In part III I have analysed the evidence for mirror matter on Earth from orthopositronium experiments and Tunguska-like explosions and summarize it now in **Figure 6.10**.

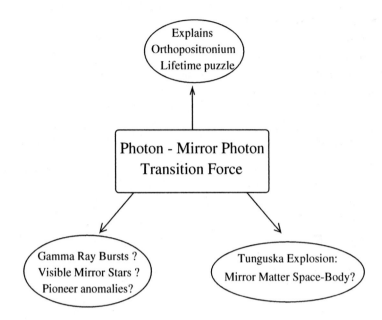

Figure 6.10: More wonders of the mirror world...

It is now time to search for mirror matter underground – which is the subject of Part IV.

I do not believe that the existence of the neutrino is definite, but do think that this hypothesis should be seriously checked or definitely disproved.

Wolfgang Pauli

Letter to O. Klein, January 8, 1931.

PART IV

Evidence for Mirror Matter from Deep Underground

Chapter 7

The Mystery of the Disappearing Neutrinos

We have already briefly encountered the neutrino. It is a very strange type of elementary particle which is produced in certain types of nuclear reactions and interacts extremely weakly. As I discussed in chapter 5, because the neutrinos are neutral (that is, they have no electric charge at all), they may therefore interact with mirror neutrinos via a microscopic transition force. Neutrinos therefore offer another window into the existence of the mirror world. Even more interesting is that the predicted effect of the mirror particles is actually found in neutrino experiments!

While the puzzles themselves are easy enough to understand, nevertheless a real appreciation requires us to delve deeper into some aspects of elementary particle physics. It is therefore time to get better acquainted with my friends the neutrinos.

A brief history of the neutrino

As I briefly discussed in chapter 5, the neutrino was first introduced to 'rescue' the law of energy conservation. It was Wolfgang Pauli who first proposed the neutrino on the 4th of December

1930, in a rather amusing letter (the most relevant and amusing parts follow) to a meeting of researchers on radioactivity in Tubingen[58]:

> Dear radioactive ladies and gentlemen,
>
> I have come upon a desperate way out regarding the 'wrong' statistics of the N- and the Li 6-nuclei, as well as to the continuous β-spectrum, in order to save the 'alternation law' of statistics and the energy law. To wit, the possibility that there could exist in the nucleus electrically neutral particles, which I shall call neutrons, which have spin 1/2 and satisfy the exclusion principle and which are further distinct from light-quanta in that they do not move with light velocity. The mass of the neutrons should be of the same order of magnitude as the electron mass and in any case not larger than 0.01 times the proton mass. The continuous β-spectrum would then become understandable from the assumption that in β-decay a neutron is emitted along with the electron, in such a way the sum of the energies of the neutron and the electron is constant.
>
> There is the further question, which forces act on the neutron? On wave mechanical grounds....the most probable model for the neutron seems to me to be that the neutron at rest is a magnetic dipole with a certain moment μ. Experiments seem to demand that the ionizing action of such a neutron cannot be bigger than that of a γ-ray, and so μ may not be larger than $e \times 10^{-13}$ cm.
>
> For the time being I dare not publish anything about this idea and address myself confidentially first to you, dear radioactive ones, with the question how it would be with the experimental proof of such a neutron, if it were to have a penetrating power equal to or about ten times larger than a γ-ray.
>
> I admit that my way out may not seem very probable *a priori* since one would probably have seen the neutrons a long time ago if they exist. But only he who dares wins, and the seriousness of the situation concerning the continuous β-spectrum is illuminated by my honoured predecessor, Mr. Debye, who recently said to me in Brussels: 'Oh, it is best not to think about this at all, as with new taxes'. One must therefore discuss seriously every road to salvation. Thus, dear radioactive ones, examine and judge. Unfortunately I cannot appear personally in Tubingen since a ball which

takes place in Zurich the night of the sixth to the seventh of December makes my presence here indispensable...

Your most humble servant, W. Pauli.

Pauli called the neutrino a neutron. At that time the 'real' neutron had not yet been discovered and Pauli hoped that his particle could lead to an understanding of the atomic nucleus[*] as well as fixing the problem of energy conservation in β-decay. For this reason Pauli's 'neutron' was a sort of mixture between what we now know to be a neutrino and a neutron. Real neutrons were discovered only about 15 months after this letter was written (that is, in 1932) which greatly clarified the structure of nuclei. It also became clear that Pauli's 'neutron' was a completely separate particle (which was later re-named 'neutrino') and over the next few years its existence began to seem more and more probable.

A reasonably successful theory for the interactions of the neutrino was developed by the Italian physicist Enrico Fermi in 1933 which greatly clarified its properties. Thus, in a short time the true nature of the neutrino emerged. The neutrino is a very strange fellow indeed. It is very light, possibly massless particle which is extremely weakly interacting. So feeble are its interactions that at its typically energy from β-decay, it would have to pass through an *average* distance of about a light year (which is 10^{13} km) of lead before it interacted just once.

Nevertheless neutrinos were detected in the laboratory during the 1950's. The experiment was done by two US Physicists, Clyde Cowan and Fred Reines, who were able to send Pauli a telegram announcing the experimental detection of the neutrino in 1956. Why? The answer is certainly not because they didn't have e-mail in 1956. The answer lies in quite a different direction. The reader may have guessed that a light year of lead was a pretty tall (long?) order. However, with the advent of nuclear reactors in the 1950's, large numbers of neutrinos were produced. The idea was if you could

---

[*]In 1930 it was widely thought that the nucleus consisted of protons and electrons, since these were the only known 'elementary' particles at the time.

produce a really large number of neutrinos, and the number of neutrinos produced from nuclear reactors was rather large – around 10^{13} per second per square centimetre, then it would be possible to eventually see a small number of them interacting.

How do you see them interacting? You build a large detector near the reactor. Inside your detector neutrinos interact with protons to produce positrons (actually the reactor was producing antineutrinos):

$$\bar{\nu} + p \rightarrow n + \bar{e}$$

The positrons ($\bar{e}$) could be detected since they would quickly annihilate with an electron producing gamma rays of characteristic energy. Neutrons could also be detected via a more subtle nuclear process. In the detector, which was essentially just water, they dissolved a chemical called cadmium chloride. The purpose of the cadmium chloride was to absorb the neutrons which are much more effective absorbers than water molecules. When the neutron is absorbed by the cadmium atom, energy is gained which is almost immediately emitted in the form of several photons (usually three or four). Furthermore, with their experimental setup the annihilation of the positron should always occur slightly before the neutron is absorbed. The time delay between these two signals – the positron annihilation and the neutron capture was only about a few millionths of a second. While this is conveniently short, it is long enough for the electronics to distinguish the two different signals.

In experiments such as this it is very important for the signal to be 'two pronged' so that background events, which have nothing to do with neutrinos, can be efficiently rejected. Such a background signal could conceivably arise from cosmic ray interactions. I need not go into the details, but such a possibility could easily be checked. The experiment can be run with the nuclear reactor turned off to see if they get any 'phantom' events masquerading as a neutrino. The result was that they found, on average about 70 more events per day with the reactor running than with the reactor shut down. In other words, 70 anti-neutrinos were absorbed and detected per day – about once every 20 minutes.

For this feat, Fred Reines was eventually awarded the Nobel

prize in 1995 nearly 40 years after the experiment (Clyde had since passed away, and Nobel prizes are not awarded posthumously).

Three neutrinos are better than one

Meanwhile, great technical advances were being made and more and more particles were being discovered during the 1950's - 1970's. This eventually led to the discovery of three 'families' of elementary (matter) particles which included not one but three neutrinos! Today's elementary matter particles are given below:

$$\begin{pmatrix} \nu_e & u \\ e & d \end{pmatrix}, \begin{pmatrix} \nu_\mu & c \\ \mu & s \end{pmatrix}, \begin{pmatrix} \nu_\tau & t \\ \tau & b \end{pmatrix},$$

$$\begin{pmatrix} \bar{\nu}_e & \bar{u} \\ \bar{e} & \bar{d} \end{pmatrix}, \begin{pmatrix} \bar{\nu}_\mu & \bar{c} \\ \bar{\mu} & \bar{s} \end{pmatrix}, \begin{pmatrix} \bar{\nu}_\tau & \bar{t} \\ \bar{\tau} & \bar{b} \end{pmatrix}.$$

Probably the only particle you recognise is the electron (e) and the electron neutrino (ν_e) and perhaps too their anti-particles which I have already mentioned from time to time. What happened to the neutrons and protons? It is now believed that these particles are actually composites of three quarks. The proton is made up of two up quarks (u) and one down quark (d), while the neutron is made of two down quarks and one up quark. Quarks can form in other combinations too – but this book is not about quarks.

Anyway, there does seem to be a lot of 'elementary' particles. That anti-particles should exist can be inferred from consistency of the quantum theory with Lorentz symmetry – as I briefly mentioned in chapter 1 – but why are there three families? In fact, nobody knows why. One family – the first one – would be enough to explain just about everything we see. The only thing that seems to distinguish the three families is their masses, otherwise their interactions appear to be completely identical. The members of the second and third families are all unstable (except for the neutrinos which appear to be completely stable) that is, they all decay into lighter particles. The first family is most familiar only because they are the lightest particles, and conservation of energy prevents them from decaying.

In addition to these particles there are also force particles which 'mediate' the three non-gravitational forces. The force particles were discussed in chapter 2 and they were summarized by the following table, reproduced here for convenience:

Force	Force particle
Electromagnetism	γ (photon)
Weak Nuclear Force	$W^{\pm}$, Z^0
Strong Nuclear Force	G^a (Gluons)

Finally the theory of all these particles – called the 'standard model of particle physics' – suggests the existence of one more particle called a 'Higgs particle'. Unfortunately, the theory does not predict the mass of the Higgs particle, so this makes it rather hard to find, even if it does exist. To date there is no confirmed sightings of this particle. The only thing that can be said is that if it exists it must be heavier than about 120 times the mass of the proton.

But, we are most interested in the neutrinos because they are neutral[*]. Experimentally, the three neutrinos are very light particles – nobody has ever succeeded in directly measuring any mass for them. Of course, this doesn't mean that they are massless, since their mass could be smaller than the sensitivity of the experimental measurements.

Interestingly, if neutrinos do have a tiny mass, then this may lead to the phenomena of neutrino oscillations. I have discussed this rather subtle quantum mechanical effect in chapter 5 in connection with orthopositronium oscillations into mirror orthopositronium. Indeed, I will discuss how the ordinary-mirror neutrino interactions (or 'mass mixing' as these 'interactions' are called in the technical literature) will cause the ordinary neutrinos to oscillate into mirror neutrinos. An effect which can be searched for in an experiment.

---

[*]The Higgs particle is also neutral. Hence, it is possible for a transition force connecting the ordinary Higgs particle with its mirror partner. If such a force exists then the properties of the Higgs particle will be modified, and this can be tested in the future if the Higgs particle is one day found in an experiment.

But permit me to let the cart go before the horse – before discussing neutrino oscillations, let us first turn our attention to the experiment. To understand this particular experiment we must first understand why the Sun shines...

Why the Sun shines and other nonsense

What makes the Sun shine?[59] In the 19th century there were only two known mechanisms to explain the source of the Sun's huge energy output. That is by gravitational contraction and chemical processes. Gravitational contraction is a process which converts gravitational potential energy into heat, in much the same way that our velocity increases when we jump out of a plane without a parachute. The problem with this mechanism is that it could explain the huge energy output for only a relatively short time, much less than 20 million years. Jumping out of a plane without a parachute is also not a viable option in the long term. It was later suggested that the Sun may have some on-going energy source from falling meteors which might help to prolong the effect. On the other hand, chemical reactions can release energy, but in insufficient quantities to sustain the huge energy output over millions of years.

Scientists, like everyone else, often have a tendency to talk a lot of nonsense from time to time (and some people and some scientists talk nonsense nearly all of the time!). Thus, because the only known sources of energy were gravitational and chemical it was decided that the Sun's energy release must be due to gravitational energy despite the fact that this leads to an age estimate much less than that suggested by geology. In 1862 the eminent scientist Lord Kelvin wrote[60]:

> That some form of the meteoric theory is certainly the true and complete explanation of solar heat can scarcely be doubted, when the following reasons are considered: (1) No other natural explanation, except by chemical action, can be conceived. (2) The chemical theory is quite insufficient, because the most energetic chemical action we know, taking place between substances amounting to the whole sun's mass, would only generate about 3,000 years' heat. (3) There

is no difficulty in accounting for 20,000,000 years' heat by
the meteoric theory.

Meanwhile, in another part of town, biologists and geologists
argued that the Sun and the Earth must have been around for much
longer than Kelvin's 20 million years. In 1859 Charles Darwin, in
his book *On the origin of the species by natural selection,* made a
crude calculation of the minimum age of the Earth by estimating
how long it would take to wash away the 'Weald' – a great valley
that stretches between the North and South Downs across the south
of England. Darwin estimated that it would take at least 300 million years to wash away the Weald. A minimum age for the Earth
also implied a minimum age for the Sun since geological evolution
required the Sun's energy. Of course, we now believe that the Earth
and the Sun are both several *billion* years old which is suggested by
radioactive dating of meteorites.

What was wrong with Lord Kelvin's logic? The short answer
is that Lord Kelvin was in 'no doubt' because he couldn't conceive
of the possibility of unknown sources of energy. This argument reminds me of some of today's scientists who believe that particles are
really tiny strings which live in ten dimensional space-time because
they argue it is the only *known* way to write down a microscopic theory of gravity... Today we know that the high central temperature of
the in-falling material that formed the Sun ignited nuclear reactions.
The basic process is the conversion of 4 hydrogen nuclei (protons)
into a helium nuclei:

$$4H \rightarrow {}^{4}He + 2\nu_e + 2\bar{e}$$

This releases a large amount of energy because in the process mass is
converted to energy via Einstein's famous relation, $E = mc^2$. A significant part of the Sun's energy is also carried off by the neutrinos
which escape from the core without interacting on the way. Thus,
the Sun is a bit like a gigantic nuclear reactor which is a source of
energy and also a source of electron neutrinos. The neutrinos coming from the Sun travel a very large distance on their way to Earth
so they could be quite sensitive to neutrino oscillations, if neutrinos
do indeed oscillate.

Actually though, Lord Kelvin's arguments were useful, even if they were nonsense. It is only by first talking a lot of nonsense that we can come by the truth. This is the way that science (and other fields) advance. This aspect of progress was rather beautifully summed up in Dostoyevsky's masterpiece *Crime and Punishment*[61]:

> 'Do you suppose I'm going on like this because they talk nonsense? Rubbish! I like it when they talk nonsense! Talking nonsense is the sole privilege mankind possesses over the other organisms. It's by talking nonsense that one gets to the truth! I talk nonsense, therefore I'm human. Not one single truth has ever been arrived at without people first having talked a dozen reams of nonsense, even ten dozen reams of it, and that's an honourable thing in its own way; well, but we can't even talk nonsense with our own brains! Talk nonsense to me, by all means, but do it with your own brain, and I shall love you for it. To talk nonsense in one's own way is almost better than to talk a truth that's someone else's; in the first instance you behave like a human being, while in the second you are merely being a parrot! The truth won't go away, but life can be knocked on the head and done in. I can think of some examples. Well, and what's our position now? We're all of us, every one of us without exception, when it comes to the fields of learning, development, thought, invention, ideals, ambition, liberalism, reason, experience and every, every, every other field you can think of, in the very lowest preparatory form of the gymnasium! We've got accustomed to making do with other people's intelligence – we're soaked in it! It's true, isn't it? Isn't what I'm saying true?....'

Of course, Razumikhin was drunk at the time – but it is nevertheless true, every word! It rather beautifully elucidates the way in which progress is made. It also touches upon the problems which can arise when people don't talk *original* nonsense, but merely 'parrot' other people's nonsense... It is only by talking *original* nonsense that one can get at the truth. I would like to write more about the importance of nonsense, but unfortunately, I do not have the literary talents of Dostoyevsky and I only know how to write nonsense. So let us return to the solar neutrinos...

The case of the missing neutrinos from the Sun

During the 1960's an experiment was proposed to detect these solar neutrinos. As I already mentioned, detecting neutrinos is not so easy because they are so elusive. The neutrinos are produced at the centre of the Sun and once produced stream outwards in all directions (at close to the speed of light, which is about 300,000 km/s). By contrast the emitted light comes from near the surface of the Sun.

If solar neutrinos are going to be useful to particle physics and the search for neutrino oscillations, then it is important to try and figure out exactly (or as precisely as possible) how many neutrinos are actually produced in the Sun. While we are at it, we might as well figure out what their energies are too. This assignment is not particularly easy. We have to work with what we know. We know the mass of the Sun (from Newton's laws) and we know its present size as well as its present luminosity. Furthermore, we also have some idea of its age. The problem of the Sun is an example of an *inverse problem*. Basically we have to build a model for the Sun, with some initial conditions, and evolve it from a ball of gas to the present day, some five billion years later. This has to be done in such a way as to reproduce its presently observed features, such as its mass, size and luminosity. In the process we can also figure out the temperature profile and the expected abundances of the various nuclei (the Sun's interior doesn't contain any atoms, it is so hot, that the electrons and nuclei are dissociated and there is only what's called a plasma of electrons and nuclei...). Knowledge of the nuclear physics then allows an estimation of the number of neutrinos and their energies.

There are various expert solar modellers who specialize in doing this type of solar modelling. The most well known is John Bahcall from Princeton University who has been in the business since the early 1960s. An important test of the solar models comes from helioseismology. The Sun doesn't just sit there. The surface undergoes slight movements in and out. Helioseismology is the study of these motions. Such studies have shown that the surface of the Sun is filled with patches that oscillate intermittently with periods of about five minutes. The detailed properties of these oscillations can provide

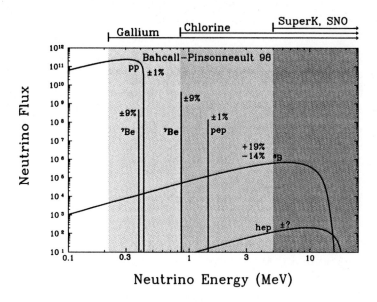

Figure 7.1: Solar neutrino spectrum. (Figure courtesy of John Bahcall's website: http://www.sns.ias.edu/ jnb/).

information on the solar interior and thereby provide a useful test of the solar model. It turns out that the standard solar model provides a reasonably accurate description of these surface oscillations which is a good reason to think that it is in good shape.

The neutrinos are produced from various nuclear reaction chains in the core of the Sun. While the main process is the conversion of hydrogen nuclei into helium nuclei, this process is only achieved by a sequence of nuclear reactions. The most important neutrino producing reactions are:

$$pp \text{ neutrinos}: \quad p + p \rightarrow {}^2H + \bar{e} + \nu_e$$
$$^7Be \text{ neutrinos}: \quad {}^7Be + e \rightarrow {}^7Li + \nu_e$$
$$^8B \text{ neutrinos}: \quad {}^8B \rightarrow {}^7Be^* + \bar{e} + \nu_e.$$

The resulting neutrino energy *spectrum* is given in **Figure 7.1**.

Every two years or so Bahcall and collaborators still update and improve their model (product?). Sometimes it goes up 10%, sometimes it goes down 10%, but otherwise it seems reasonably robust.

The first experiment to get underway started taking regular data in 1970 and ran almost continuously into the 1990's. The experimental team was led by Ray Davis – another neutrino pioneer. This experiment was located deep underground in the Homestake gold mine in South Dakota. It consisted of a large tank containing 615 tons of tetrachloroethylene – a type of cleaning fluid. The idea was that the neutrinos would occasionally convert the chlorine in the cleaning fluid into Argon, via the weak interaction process:

$$\nu_e + {}^{37}Cl \rightarrow e + {}^{37}Ar$$

Argon is a gas and it could be extracted and counted.

Looking for Argon atoms in cleaning fluid several kilometres underground in a gold mine doesn't sound much like fun. Nevertheless, one can always look for gold between atom counting! Anyway, by counting the number of Argon atoms they could figure out the number of neutrino interactions in their detector and compare this with the number that they expected. The expected number of neutrino interactions is simply a product of the number of neutrinos passing through their detector multiplied by the probability that they interact (which is a very large number times a very small number). The number passing through their detector can easily be related to the predicted number coming from the Sun. The probability that the solar neutrinos will interact within their detector is now known reasonably accurately. Of course, modelling the Sun is not so easy and there are a number of uncertainties coming from various uncertain nuclear reactions and approximations.

At the end of the day, or in this case, at the end of nearly 30 years of pumping out Argon, the result is that the measured number of neutrinos is much less than the number they had expected. It is about one third of the expected value. If the measured value had been within about 20% or even 30% of the predicted value then this would have been okay. Such a discrepancy could just be explained by the various uncertainties in the solar model and experiment. However, the measured value was only about a third of the predicted value. This puzzle has been called the *solar neutrino puzzle* and has been a mystery for a long time – several decades in fact.

Way back in 1968 it was proposed[62] by one Russian and one Italian (who defected to Russia in 1950) – Gribov and Pontecorvo – that this discrepancy might be explained by neutrino oscillations. So it is now time to delve deeper into this weird quantum mechanical effect.

Do leopards change their spots?

In quantum theory each particle is described by a wave function, Ψ, which contains all the information about the particle. For example, knowledge of Ψ tells us the energy, momentum, spin, charge, mass etc. In biology they have a similar thing called DNA. DNA encodes all of the properties of each living organism. The wave function is also called a 'quantum state'. In quantum theory linear combinations of wave functions are also possible quantum states. However, I suspect that linear combinations of DNA strands may not always lead to a living organism. But I must confess, I don't know much about biology. So let us stick to wave functions. One could imagine forming a linear combination of two states with different masses. That is, if Ψ_1 is a quantum state describing a particle with mass m_1 and Ψ_2 is a quantum state describing a particle with mass m_2, then

$$\Psi_{new} = \cos\theta\Psi_1 + \sin\theta\Psi_2$$

is a possible quantum state.

In quantum mechanics, the coefficients of the component wave functions are related to the probability. In the particular case described by the state Ψ_{new}, the state has a $\cos^2\theta$ probability of behaving like a particle with mass m_1 and a $\sin^2\theta$ probability of behaving like a particle with mass m_2. Of course, there are an infinite number of such combinations, one for every value of θ. In practice though, θ has a fixed value which ultimately depends on the fundamental (microscopic) theory.

A strange quantum mechanical effect called 'particle oscillations' can occur if one does indeed produce a state which is a linear combination of two mass states. We briefly encountered this phenomenon in chapter 5 in our discussions of orthopositronium.

What happens is that the states Ψ_1 and Ψ_2 evolve differently in time. Why? Because states with different masses have different energies, and in quantum theory – there is a profound connection between energy and time. In chapter 5 I already mentioned that energy conservation was related to time translational symmetry. Anyway, what happens is that Ψ_1 and Ψ_2 evolve in time, each in a different way, so that after a while the linear combination has changed to become a different linear combination compared to when the particle was produced. A different linear combination is, in essence, a different particle. The identity of the particle literally oscillates as it travels along. This is a very strange effect. After all, our grandmothers told us that leopards never change their spots – but perhaps there is a difference between leopards and particles.

Not all particles can change their identity by oscillating. There are certain conditions that must be satisfied. For example, the mass difference between the states must be much less than the particle's energy. Thus, for instance, we couldn't get the u-quark to oscillate into a c-quark because there is a large difference in their masses at their typical energies where they are produced.

Do the neutrinos oscillate?

It is time now to turn our gaze onto our neutrino friends. Do they oscillate and change their identity as they travel along? First, the neutrinos are known to be very light so that their mass difference is less than their typical production energy – which is one necessary condition for oscillations to occur. We now know that there are three different types of neutrino, ν_e, ν_μ, ν_τ (and their anti-particles). These three states have very particular interaction properties. For example, the ν_e is always produced along with a positron ($\bar{e}$) when it interacts with a W^+ force particle. In contrast, the ν_μ is produced along with a muon ($\bar{\mu}$), and the ν_τ is produced along with a tauon ($\bar{\tau}$) when interacting with a W^+ force particle. However, it is a possibility that Nature may not make these interaction states to be simultaneously states of definite mass. In other words, these interaction states might be some combination of two or more states of definite mass. As I discussed above, this means that if we produce an electron neutrino, ν_e, from β-decay, then over time the neutrino evolves so that at a later time it can behave like a ν_μ or ν_τ state. In fact the identity of the state actually oscillates in time. But, the neutrinos are travelling at close to the speed of light. Thus, the identity of the neutrino oscillates as they travel over some distance. In an experiment, the relevant distance is the distance between the production point and the detection point. As I mentioned in chapter 5, this magician's trick has been observed in the laboratory for kaons, there is some evidence that it is occurring for orthopositronium and it is close to being established that it also occurs for neutrinos.

Let us consider a specific illustrative example. Imagine that there are only two types of neutrinos, ν_e, ν_μ. Imagine further that each of these interaction states are indeed not simultaneously states of definite mass. Instead they are a certain mixture of mass states as I described earlier (in terms of Ψ_1, Ψ_2). Such mixing is described by a 'mixing angle', θ. This angle can be viewed as a parameter of the fundamental theory just like the electron mass or the electron charge. Particle masses and parameters like θ are essentially free parameters which 'nature' fixes to some particular value. They are part of Newton's 'great ocean of truth' that still hasn't been uncovered yet. Put another way, there is no compelling theory which explains their precise values.

In quantum theory, the electron neutrinos and muon neutrinos are then expressed in terms of the mass states, which mathematically take the form:

$$\nu_e = \sin\theta\Psi_1 + \cos\theta\Psi_2, \ \nu_\mu = \cos\theta\Psi_1 - \sin\theta\Psi_2.$$

We need not be too concerned though with mathematical details. The important point is that because Ψ_1 and Ψ_2 evolve in time (or distance) differently, an initial ν_e state could end up looking a bit like a ν_μ state. Quantum mechanics allows us to write down the probability that an initial ν_e state will behave like a ν_μ state after travelling a distance L. It is given by

$$P(\nu_e \to \nu_\mu) = \sin^2 2\theta \sin^2 (L/L_{osc}).$$

The quantity L_{osc} is the oscillation length. The oscillation length depends on the neutrino energy and the neutrino masses[*]. From the above equation we can see that this transition probability is proportional to the square of the quantity $\sin 2\theta$. From our high school days we all learned that $\sin 2\theta = 2 \sin \theta \cos \theta$. Evidently, the transition probability vanishes if $\sin \theta = 0$, while if $\sin 2\theta = 1$ the oscillations have the greatest effect and this is called 'maximal oscillations'.

Clearly, the mixing angle θ is very important because it strongly influences the 'size' of the oscillation effect. There are several reasons for suspecting that θ (and hence $\sin 2\theta$) is small. None of them are rigorous in the sense that it must be small. Mathematically θ can be anything and things are still consistent. But physics is not mathematics. Nature chooses some value for θ and we have to try and figure out what it is.

Generally the masses of the charged particles in each family are well separated. [Recall that the three 'families' of elementary particles were briefly discussed on page 179]. The average mass of the first family is roughly a few MeV, the average mass of the second family is roughly a few hundred MeV while the average mass of the third family is roughly 10,000 MeV or 10 GeV (1 GeV = 1000 MeV). The alert reader may have noticed that our units correspond to energy not mass. As already mentioned in chapter 2, particle physicists tend to use mass and energy almost interchangeably because of the mass-energy equivalence, $E = mc^2$. Anyway, the point is that the large mass separation between each of the three families of elementary particles does suggest that they do not mix together much. Indeed, in the case of quarks, the mixing angles are all measured and are found to be small. Thus, in many ways small mixing between the three families appears to be the most natural possibility. If θ is in fact small, then the effect of oscillations will not be large since the oscillation probability is proportional to $\sin 2\theta$ and $\sin 2\theta \to 0$ if $\theta \to 0$. This means that we could expect that only a

---

[*]Technically, the 'real' oscillation length is not L_{osc} but $\pi \times L_{osc}$, however this detail is unimportant. For people interested in more technical details, let me also add that the precise way in which L_{osc} depends on the neutrino masses and the energy is $L_{osc} \approx 0.80 E/\delta m^2$, where L_{osc} is in units of metres, the neutrino energy, E, is in units of MeV, and $\delta m^2 = m_1^2 - m_2^2$ is the difference in mass squared of the Ψ_1 and Ψ_2 states (in units of eV^2).

small fraction of electron neutrinos produced within the Sun can be converted into muon or tauon neutrinos from the oscillation effect. If θ is small between families, what about between ordinary and mirror neutrinos? Could oscillations between say ν_e and ν'_e be large?

Adding the mirror neutrinos

In 1991 Henry Lew, Ray Volkas and I began studying the effect of the ordinary neutrino-mirror neutrino transition force or 'mass mixing' force. The main observable consequence of this 'mass mixing' is that it leads to oscillations of ordinary neutrinos into mirror neutrinos, rather than of oscillations between neutrinos of different families (although oscillations between ν_e, ν_μ and ν_τ may also occur). Anyway, when we examined the effects of this possible force, we immediately found something rather interesting. We found that the oscillations between ordinary and mirror neutrinos are *necessarily* large. In fact, we found that $\sin 2\theta = 1$, which is maximal mixing! This result was a direct consequence of the mirror symmetry. It is difficult to explain this result at an elementary level, but I will try to make it plausible.

In quantum theory a special role is played by something called an 'eigenstate'. An eigenstate is something which doesn't change its identity under a symmetry. For example, under the mirror reflection symmetry, the ordinary electron neutrino (ν_e) is reflected into a completely new particle, the mirror electron neutrino (ν'_e), that is, $\nu_e \leftrightarrow \nu'_e$. This means that ν_e and ν'_e are *not* mirror symmetry eigenstates because they change their identity under the mirror symmetry. Instead of ν_e and ν'_e we could consider the particular combinations defined by $\nu^+ = \nu_e + \nu'_e$ and $\nu^- = \nu_e - \nu'_e$. Then under the mirror symmetry $\nu^+ \to \nu^+$ and $\nu^- \to -\nu^-$. Evidently, these particular combinations of states do not change their identity when reflected in the mirror. These states are called states of definite 'mirrorness' or 'parity'. That is they are the 'eigenstates' of parity. The ν^+ combination has positive parity (because $\nu^+ \to +\nu^+$ under the mirror reflection) while the ν^- state has negative parity (because $\nu^- \to -\nu^-$ under mirror reflection).

A consequence of the parity symmetry is that the states of definite parity ($\nu^{\pm}$) are always states of definite mass (or energy). If the combinations, $\nu^{+} = \nu_e + \nu'_e$ and $\nu^{-} = \nu_e - \nu'_e$ are the states of definite mass, then it follows, by adding and subtracting ν^{+} and ν^{-}, that

$$\nu_e = \nu^{+} + \nu^{-}, \quad \nu'_e = \nu^{+} - \nu^{-}. \tag{*}$$

Actually the equal sign ('=') should really be a proportional sign ('$\propto$') since I have dropped a 'normalization factor' – but don't worry about that! The important point is that the above particular combinations are very special. Recall that in the most general case (for $\nu_e \to \nu_\mu$ oscillations) discussed earlier, we had

$$\nu_e = \sin\theta \Psi_1 + \cos\theta \Psi_2, \quad \nu_\mu = \cos\theta \Psi_1 - \sin\theta \Psi_2$$

where θ is a theoretically unknown parameter. However, for oscillations of $\nu_e \to \nu'_e$, the mixing angle, θ is *not* a free parameter at all. Comparing the above equation with the corresponding one describing the states ν_e and ν'_e [Eq.(*)] we see that $\sin\theta = \cos\theta$ which means that $\theta = 45°$ i.e. $\sin 2\theta = 1$ and hence maximal mixing! Thus, if neutrinos and mirror neutrinos have mass and mix together, the oscillations between the ordinary and mirror neutrinos are necessarily maximal. This is interesting because it means that the effect of oscillations can be very large. Maybe large enough to explain the missing solar neutrinos.

The point is that the neutrino experiments which detect neutrinos from the Sun, such as the Homestake experiment discussed earlier, are usually only sensitive to electron neutrinos and these are the only neutrinos produced in the Sun. If the electron neutrinos oscillate into something else, for example a muon neutrino or perhaps a mirror electron neutrino, then the measured number would really be less because a fraction of the electron neutrinos have transformed themselves into something else by the time they reached the detector. However, to explain the large discrepancy between the experiment and the theory clearly suggests that the oscillation effect must be significant; something like a half or two thirds of the neutrinos have to oscillate into something else. Clearly, a large oscillation effect is required. The largest effect occurs for maximal oscillations, where the oscillation probability is greatest.

The reader may have noticed the sharp contrast to orthopositronium oscillations discussed in chapter 5. Recall that for orthopositronium we found a very small oscillation effect. Orthopositronium oscillations are also maximal because this is a general feature of oscillations of particles into mirror particles. How can the oscillation effect be small for orthopositronium and large for solar neutrinos? The answer is quite simple. The difference is that orthopositronium is very short-lived and decays in such a short time so that only a fraction of an oscillation has occurred. For this reason the oscillation effect is therefore quite small for orthopositronium. Neutrinos on the other hand, are either completely stable or very long-lived. Nobody has ever observed one to decay. Furthermore, it takes solar neutrinos about eight minutes to travel the 150 million kilometres from the Sun to the Earth. Provided that the neutrino oscillation length, L_{osc}, is less than or of the order of 150 million kilometres, the oscillations can lead to a very large effect.

A useful limit is when the oscillation length is much smaller than the distance between the Earth and the Sun. In this case the effect of oscillations is to average the oscillation probability. This is because of the range of distances varies slightly since the neutrinos are produced over a finite region in the Sun. Also, they have a range of energies which to some extent is averaged out in an experimental measurement. Finally, for a range of oscillation lengths certain 'matter effects' occur which must also be taken into account and can lead to important effects which will be discussed in more detail in a moment. The net effect though, is that to a good approximation, $\sin^2(L/L_{osc})$ can be replaced by its average value of 1/2 and the oscillation probability is then given simply by

$$P(\nu_e \to \nu_X) = \frac{1}{2}\sin^2 2\theta$$

It follows that for maximal oscillations (that is $\sin 2\theta = 1$) of ν_e into ν_e' predicted by the mirror matter theory, we have simply that the oscillation probability is 1/2. In other words, precisely half of the electron neutrinos would be transformed into mirror neutrinos on the way from the Sun to the Earth. Mirror neutrinos don't interact at all with an ordinary matter neutrino detector, which effectively means that half of the solar neutrinos disappear! This doesn't give a perfect fit to the Homestake experiment which measures only about a third of the expected number instead of a half, but no experiment is perfect!

More solar neutrino experiments are done

Since 1991 there has been a lot of experimental progress on solar neutrinos. There are now no less than seven experiments (Homestake and six new ones) which have already published data, and several more very important experiments either currently taking data or about to be switched on for the first time. The interesting thing is that they all found approximately half of the number of neutrinos that they expected, only the first experiment (Homestake) found a third. This doesn't necessarily mean that Homestake is wrong since the other experiments measure neutrinos from different regions of the energy spectrum. However, common sense suggests that the number of neutrinos is probably reduced by about a half, and the difference between the experiments is most likely due to experimental or theoretical uncertainties.

As I have mentioned above, electron neutrinos emitted from the Sun arise from various nuclear reactions in the solar core. Theoretically the most important are the pp reaction chain where two protons fuse together to form deuterium: $p + p \rightarrow \ ^2H + \bar{e} + \nu_e$. This neutrino flux can be most reliably predicted since it is directly related to the luminosity of the Sun. There are three experiments specifically designed to measure these pp neutrinos which are called SAGE, GALLEX and GNO. The SAGE and GALLEX experiments began taking data around 1991 with GNO starting in 1998. Their most recent experimental results divided by the theoretical prediction are

$$0.52 \ \pm \ 0.08 \ (SAGE),$$
$$0.59 \ \pm \ 0.08 \ (GALLEX),$$
$$0.50 \ \pm \ 0.10 \ (GNO),$$

where the uncertainties contain the estimated experimental *and* theoretical errors. These numbers should first be compared with 1.0,

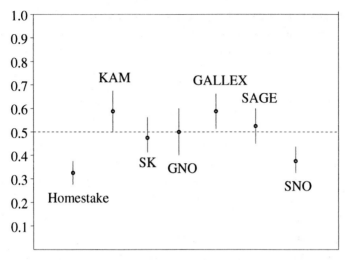

Figure 7.2: Summary of the experimental measurements divided by the theoretically expected value – assuming that no oscillations are occurring. The dotted line is the expected value if ordinary neutrinos oscillate into mirror neutrinos.

the expected value if neutrino oscillations do not occur, or 0.5 if oscillations into mirror neutrinos do occur (with an oscillation length much less than the Earth-Sun distance). Clearly, the above measurements are consistent with the mirror matter prediction of 0.50. As I explained in chapter 5, if the theory and experiment results are within about twice the quoted error then the agreement is generally considered to be quite good. [The quoted error actually means that there is only about a two thirds probability that the answer is within the uncertainty range].

More recently, the Super-Kamiokande Collaboration have reported an energy independent (within errors) recoil electron energy spectrum in their experiment designed to measure the ^8B neutrinos (i.e. neutrinos from the nuclear reaction, $^8B \rightarrow {}^7Be + \bar{e} + \nu_e$), also finding only 50% of the expected solar flux. Again these results were predicted in the mirror matter theory.

The current status of the solar neutrino experiments are summarized in **Figure 7.2** (above).

Much ado about nothing?

The totality of the results from the solar neutrino experiments does strongly support the idea that half of the solar neutrinos (which are electron neutrinos) have disappeared. This disappearance is consistent with maximal electron neutrino-mirror electron neutrino oscillations predicted by the mirror matter theory way back in 1991.

Still, more work needs to be done. It needs to be shown that oscillations are really happening. Although the current evidence for solar neutrino oscillations is very strong, it is not yet completely compelling. For example, it is possible to imagine that John Bahcall and his friends have made some mistakes in their calculations of the neutrino flux from the Sun. Maybe some of their assumptions don't work. Nobody is perfect. Of course, there is no indication for this and there is evidence (for example from helioseismology) that John has done a very good job. I personally believe that the solar neutrino problem does exist and is solved by maximal electron neutrino oscillations (although it may be that ν_e oscillates into ν_μ or ν_τ rather than ν_e'). But, as Pauli said, 'Between believing and knowing is a difference and in the last end such questions must be decided experimentally'. If oscillations are occurring then their tell-tale effects must show up in two new experiments called 'KamLAND' and 'Borexino'.

The KamLAND experiment is run by an international collaboration which includes Hungary, Japan and the US. This collaboration is currently constructing a large neutrino detector in the underground site that used to be the home of the Kamiokande experiment. Called KamLAND (Kamioka Liquid scintillator Anti-Neutrino Detector), it will be the largest low-energy antineutrino detector ever built. The detector is located roughly in the middle of Japan's main island of Honshu. Its primary purpose is not to detect neutrinos from the Sun, but to detect neutrinos from several commercial nuclear reactors at a distance of between 150 and 200 kilometres. KamLAND will be able to discover neutrino oscillations provided that the oscillation length is less than that distance (200 km) for their typical energy of about a million electron Volts (MeV). Furthermore, because the oscillation length depends on the neutrino energy in a known way, KamLAND should find a certain distortion to their neutrino energy spectrum. Such a distortion does not show up in the solar neutrino experiments because of the very large distance involved which means that all the effects of oscillations have averaged out. Previous, but much smaller scale reactor experiments did not find any evidence for oscillations, which allows a lower limit on the oscillation length of electron neutrinos. The oscillation length, which depends on the neutrino masses and neutrino energies, can be thought of as a free parameter since theoretically the neutrino masses are unknown. We do know that neutrinos are very light, but we don't know exactly how light they are. Anyway, for neutrino energies of about an MeV, the previous reactor experiments[*] tell us that the neutrino oscillation length is greater than about one kilometre. However, to solve the solar neutrino problem the oscillation length must be less than the distance between the Earth and the Sun for all observable energies. This translates into an upper limit for the oscillation length for 1 MeV neutrinos of about three million kilometres.

In summary, if

$$1 \text{ km} < L_{osc} < 3,000,000 \text{ km}$$

then the maximal oscillations will reduce the flux of solar electron neutrinos by approximately 50% which is broadly consistent with the laboratory measurements.

Day-night effect

There is a subtle effect which sometimes happens when neutrinos oscillate in matter. As the neutrino propagates, it occasionally passes near nuclei and feels small potential energies arising from the weak nuclear force. It turns out that this effect can potentially modify the neutrino oscillation rate. This 'matter effect' as it is called,

[*]The most sensitive such experiment is the CHOOZ reactor experiment in the Ardennes region of France.

can therefore be important for solar neutrinos since these neutrinos go through matter during their outward passage through the Sun and also during their passage through the Earth. While day-time neutrinos travel only a few kilometres though the Earth before they reach the underground detector, night-time neutrinos have to go all the way through the Earth (typically several thousand kilometres) before they arrive at the detector. The most important observable consequence of this matter effect is that it can make the night-time rate slightly greater than the day-time rate. In other words, the Sun appears to shine brighter in neutrinos at night-time than in the day-time. However, this effect only happens if the neutrino oscillation length happens to be in a particular range.

The only current experiment which can search for this 'day-night effect' is the super-Kamiokande experiment. This is because super-Kamiokande is a 'real time experiment'. Neutrino events are recorded as they occur which obviously allows night and day events to be distinguished. In the other experiments such as Homestake, the data is sampled only once every month or so which means the night and day events get added together and cannot be separately counted. Anyway, super-Kamiokande have searched for a day-night effect, but have not found any difference between their day-time and night-time rates. The fact that they didn't find any effect allows a slice of parameter space to be excluded. However, a new experiment called Borexino will probe a different range of neutrino energies and should observe a day-night effect if the neutrino oscillation length is very long. The current and near future situation is summarized in **Figure 7.3**.

SNO news is good news

A very important issue is to experimentally distinguish $\nu_e \rightarrow \nu_e'$ oscillations from $\nu_e \rightarrow \nu_\alpha$ where $\nu_\alpha = \nu_\mu, \nu_\tau$. This can be done in a new experiment at the Sudbury Neutrino Observatory (SNO). SNO is located about two kilometres underground in the Creighton mine near Sudbury, Ontario (Canada). The SNO neutrino detector consists of 1000 tonnes of heavy water on loan from Atomic Energy

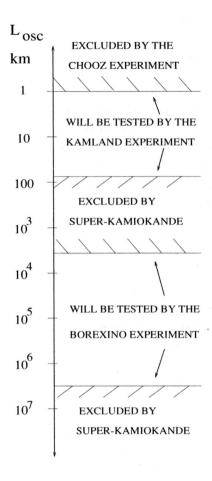

Figure 7.3: Summary of what we know about the oscillation length for solar neutrinos, L_{osc} (here defined for neutrinos of energy 1 MeV). If the oscillation length were very short, $L_{osc} < 1$ km then it would have shown up in the CHOOZ reactor experiment. A slice 80 km $< L_{osc} < 4000$ km can also be excluded since it leads to a day-night difference which is also *not* seen at super-Kamiokande. Finally, if the oscillation length were very long, $L_{osc} > 3$ million km, then energy *dependent* distortion should occur which is *not* seen by the super-Kamiokande experiment (and if $L_{osc} \gg$ 100 million km then there would be no oscillation effect at all because the neutrinos would not have time to oscillate on their way from the Sun to the Earth).

Canada Ltd. Ordinary water molecules are composed of two hydrogen atoms and one oxygen atom – H_2O. Heavy water (D_2O) is similar except the hydrogen atoms are replaced with the isotope – deuterium (D). Deuterium is like hydrogen except that its nucleus contains one proton and one neutron (while hydrogen's nucleus consists of just a solitary proton). Because of the extra neutron, heavy water is about 10% heavier than ordinary water, but otherwise tastes the same. It is true though, that it is very expensive to make so it's best not to drink it. The heavy water is contained in an acrylic vessel with a diameter of 12 metres. Around this vessel is a support structure which holds 9600 photomultiplier tubes. The acrylic vessel and support structure is itself immersed in normal water within a 30 metre barrel-shaped cavity two kilometres underground. [See **Figure 7.4** opposite, for an illustration of the SNO detector].

Needless to say that excavating such a large cavity and building such a big detector was not an easy task. This experiment was conceived about 20 years ago. A collaboration was set up in 1984, which today involves more than 100 scientists from Canada, UK and the US. Following a feasibility study in 1985 a technical proposal was written, reviewed and eventually things got going. In March 1990 excavation of the cavity began and was finished about three years later. Then the laboratory was constructed in the cavity, the acrylic vessel was made as were 10,000 phototubes, the electronics, the water purification system etc. After all that, the calibration of the detector could begin... Finally in November 1999 data taking began with the first results just released in June 2001.

How does SNO detect neutrinos? This detector can detect neutrinos in several different ways. Firstly, it can detect electron neutrinos, ν_e, via the weak *charged current* process. In this process an electron neutrino interacts with the deuterium (D) to produce two protons and an electron:

$$\nu_e + D \rightarrow p + p + e$$

The electron is produced with enough energy to travel faster than the speed of light in water. [This is not in conflict with Einstein's relativity theory because the speed of the electron is still less than the speed of light in vacuum]. This causes the optical equivalent

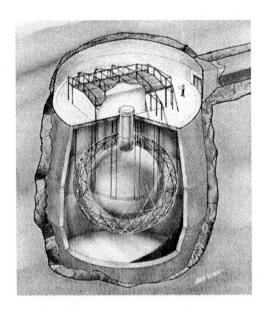

Figure 7.4: The SNO detector. (Figure courtesy of the SNO website, http://www.sno.phy.queensu.ca).

of a "sonic boom" where a "shockwave of light" is emitted by the electron as it slows down. The flash of light – called Cherenkov radiation – is collected by the photomultiplier tubes surrounding the heavy water vessel. From the properties of the light the energy and direction of the electron can be determined, which allows information about the 'parent' electron neutrino to be obtained. For this charged current process, the number of electrons that is expected to be detected has been estimated to be about 30 per day.

Another completely different way to detect solar neutrinos comes from the *neutral current* process:

$$\nu_x + D \rightarrow p + n + \nu_x$$

where $\nu_x = \nu_e, \nu_\mu$ or ν_τ. In this case no electron is produced, however the neutron in deuterium is liberated. If this neutron is captured it can emit several gamma rays, as was discussed earlier in relation to the Cowan-Reines experiment which first detected the neutrino. However, neutrons are not easily captured in water. To improve the sensitivity Chlorine can be added in the form of salt - NaCl, which is much better at capturing neutrons.

The important thing about this neutral current process is that it is sensitive to all three of the known neutrinos, ν_e, ν_μ, ν_τ, not just ν_e. Thus, if half of the ν_e oscillates into ν_μ or ν_τ then the rate for this process will *not* change, while if ν_e oscillates into ν_e' then the neutral current rate will be 50% lower. In both cases though, the charged current process will be 50% lower since that process is only sensitive to ν_e. The first SNO NC results are expected to be announced sometime during 2002.

A third distinct way in which the SNO experiment can detect solar neutrinos is by the elastic scattering of neutrinos off electrons [in the water molecules], which has a contribution from both charged current and neutral current processes. This is the same process by which the super-Kamiokande experiment detects solar neutrinos. If one combines the super-Kamiokande electron elastic scattering measurement with the recent SNO charged current measurement, then it is possible to extract a rough estimate of the neutral current rate. This neutral current estimate does *not* currently favour oscillations of ν_e into its mirror partner. This is an interesting hint *against* the mirror neutrino oscillation hypothesis for solving the solar neutrino problem, but is not accurate enough at the moment to be really convincing. Still, it does hint that SNO news may not be good news... Things should become clearer in the near future when SNO finally measures *directly* the neutral current rate.

Of course, I should perhaps mention that although mirror symmetry tells us that the oscillations of neutrinos into mirror neutrinos are maximal, it does not tell us anything about the size of their oscillation length, L_{osc}. In the limit whereby neutrino-mirror neutrino mass mixing goes to zero, any effect of the mirror world must also vanish. This means that the oscillation length for $\nu_e \to \nu_e'$ oscillations becomes infinitely large in that limit. Furthermore, the existence or non-existence of a mirror world does not tell us anything about the oscillations between the three known neutrinos ν_e, ν_μ, ν_τ.

I have simply assumed that there are not large oscillation effects between ν_e and ν_μ, ν_τ, however this assumption could be wrong. Obviously, if experiments do happen to prove that the disappearance of solar neutrinos are due to $\nu_e \rightarrow \nu_\mu, \nu_\tau$ oscillations rather than $\nu_e \rightarrow \nu_e'$ oscillations, then it would mean that the oscillation length of $\nu_e \rightarrow \nu_e'$ oscillations must be greater than the Earth-Sun distance and also that our assumption of small effects between the three known neutrinos is wrong. Still, it could be viewed as another 'great tragedy of science: the slaying of a beautiful hypothesis by an ugly fact'. The hypothesis in this case being that the solar neutrino problem is solved simply by maximal $\nu_e \rightarrow \nu_e'$ oscillations...

A great mathematician (David Hilbert), once said, 'Physics is much too hard for Physicists'. Real progress in physics requires a number of important ingredients. First, we must come up with interesting, plausible, and logically consistent ideas which lead to experimentally testable consequences. Second, we must make sure that they agree with all of the existing experiments. The third requirement, though, is the most challenging of all. Our ideas, if they are to describe nature, must agree with new experiments as they are done. This last requirement assumes of course that the experiments are free from mistakes. In science things often become complicated when experiments are slow, difficult, and expensive as they are in neutrino physics, because this makes it difficult for them to be carefully checked independently. Nevertheless, there is reason to believe that eventually things should get sorted out...

Chapter 8

More Missing Neutrinos

In the previous chapter I have discussed the mystery of the missing neutrinos from the Sun. In that case, the number of missing neutrinos was about half the expected number and that this factor of a half could be explained in the mirror matter theory by the oscillations of ordinary neutrinos into mirror neutrinos. I also pointed out that there are actually three distinct types of neutrinos called electron, muon and tauon neutrinos (and three corresponding anti-neutrinos). The Sun is a source of electron neutrinos only.

In the mirror matter theory we would expect that each of these three neutrinos should oscillate into their mirror partner – not just the electron neutrino. Is there any evidence for missing muon or tauon neutrinos? While at the present time, there is essentially no experimental tests on the possibility of tauon neutrino oscillations, there is a lot of evidence that muon neutrinos oscillate. The evidence is even stronger than the evidence for electron neutrino oscillations from solar neutrino experiments. In fact, the evidence is widely considered to be compelling. What's not so clear though, is the identity of the oscillating partner of the muon neutrino. In other words, we have a smoking gun but the owner of the gun has fled the scene of the crime.

Sometimes the God of physics works in mysterious ways. The case of the disappearing muon neutrinos is an example of this.

Grand Unification and proton decay

The story begins in the 1970's. During the 1970's it eventually became clear that the three non-gravitational forces – electromagnetism, weak and strong nuclear forces – could all be described in terms of the exchange of force particles. The interactions of these force particles with themselves and with matter particles were expressed by a mathematical theory which exhibited a large variety of symmetries. However, the force particles for each force were still a bit different and it was argued that there might exist some type of 'higher symmetry' under which the force particles would be related to each other, just as the different directions of space are related by rotations.

Such a theory was quickly invented, and it did seem very interesting and really possible. In short – it was a very good idea. The simplest theory was written down by Howard Georgi and Shelly Glashow in 1973. The most memorable part of their paper was the sentence[63]:

> Our hypothesis may be wrong and our speculations idle, but
> the uniqueness and simplicity of our scheme are reasons enough
> that it be taken seriously.

In fact, it was taken seriously. Its main testable prediction was that protons should be unstable and decay into a positron and a short lived particle called a π^0 (as well as various other decay modes): *

$$p \to \bar{e} + \pi^0$$

Of course the rate at which the proton decays must be very low since otherwise we would all decay. Not just tooth decay, but all of the parts of our bodies are made of atoms – which contain protons. If

---

*Previously (in chapter 2) it was mentioned that protons *within* certain nuclei can decay via the weak interaction. Nevertheless in isolation as well as in most nuclei they are completely stable: energy conservation prevents them from decaying via the weak nuclear force. In grand unified theories (GUT) protons decay via a *new* type of interaction which would cause even isolated protons (as well as protons in all nuclei) to decay into light particles, $\bar{e}$, π^0 (which is not forbidden by energy conservation).

protons are spontaneously decaying then we would be slowly disintegrating. Of course, not just us, but the Earth, the Moon and the stars. This might not be a problem for the theory, though, if the average lifetime of the proton was very very long.

Within a few years, the average lifetime of the proton was estimated within the theory. It was estimated to be about 10^{31} years. This is indeed a very very long time, so there were no apparent problems with the longevity of ordinary matter. It is even consistent with the big bang model of the Universe which suggests that the Universe itself is only about 15 billion years old (give or take a billion years). People liked the theory so much they decided to give it the glorious title of 'Grand Unified Theory', also known by the somewhat less glorious acronym 'GUTs'. Conferences were held every year and there was great joy in the land. It was proclaimed that Georgi and Glashow had succeeded while Einstein had failed in such a quest several decades earlier[*].

Of course 10^{31} years is a long time to wait for a proton to decay. However, if you got together 10^{31} protons (about the number in a small swimming pool), then one would decay each year. And that's more or less what they did. Large tanks of water were constructed containing about 10^{33} protons. They had to construct large tanks of water because ordinary swimming pools were no good because of the background effects from swimmers and cosmic rays. While swimmers could be eliminated, cosmic ray effects were more troublesome. Cosmic rays could produce lots of positrons (and other particles) in the detector which would make it difficult to distinguish from a real proton decay. One cannot 'turn off' proton decay to measure the background rate, and then turn it on again. So, the swimming pool had to be an 'in-ground pool'. In fact, it was an 'underground pool'. In order to reduce the cosmic ray background, the experiment was constructed in large underground facilities (usually in existing mines) at depths of more than one kilometre below the

---

[*]It was even argued that GUTs could explain 'electric charge quantization', the observed fact that the electric charge of the proton appears to exactly cancel the electric charge of the electron. Later it was realised though, that electric charge quantization could be simply understood without GUTs provided that the neutrinos had mass.

surface. The same facilities were also used to detect solar neutrinos described in the previous chapter. It was real value for money. Two interesting experiments for the price of one.

While putting the detector deep underground was a great help in reducing the background from cosmic rays, it did not eliminate them completely, since the neutrinos produced in the atmosphere can easily penetrate one kilometre of earth. Furthermore, atmospheric neutrinos are quite different from solar neutrinos since they are much more energetic. Atmospheric neutrinos have enough energy to produce an energetic positron and π^0 with enough energy to mimic a decaying proton – assuming protons do indeed decay. The process is the capture of an electron anti-neutrino by a proton to produce a positron, a π^0 and a neutron:

$$\bar{\nu}_e + p \to \bar{e} + \pi^0 + n$$

Experimentally the positron and π^0 produced by atmospheric neutrino interactions in the water tank could be indistinguishable from a proton decay positron and π^0. So a lot of studies were done trying to understand atmospheric neutrinos.

From the point of view of the GUT enthusiasts the cosmic ray induced neutrino events[64] 'produce an undesirable and confusing background in the yet inconclusive search for proton decay.' However, from the point of view of neutrino oscillation enthusiasts and especially mirror matter enthusiasts, these 'background' neutrino events were much more interesting than the potential proton decay signal. So it is time to become better acquainted with these atmospheric neutrino events.

Atmospheric neutrinos

The atmosphere of our planet is continually bombarded by cosmic rays. These are mostly high energy protons which originate from distant stars (or stellar remnants). When these high energy protons strike the atmosphere they collide with the molecules of the air, that is, nitrogen and oxygen. The resulting products of such collisions often include a host of short-lived particles called muons (μ), pions (π) and kaons (K). Most of these particles eventually decay while still in the atmosphere into electron and muon neutrinos and their anti-neutrinos. In summary we have the interaction chain:

$$p + air \rightarrow \mu, \pi, K + \text{other stuff}$$
$$\rightarrow \nu_e, \bar{\nu}_e, \nu_\mu, \bar{\nu}_\mu + \text{other stuff.}$$

'Other stuff' includes left-over protons and electrons and also some short-lived particles such as muons which have not yet decayed. Most of the short-lived charged particles which do not get time to decay during their atmospheric flight are stopped in the Earth before reaching the underground detector.

A small fraction of the neutrinos can interact in the detector to produce an electron, positron, muon or anti-muon, via the weak interaction processes:

$$\nu_e + n \rightarrow e + p$$
$$\bar{\nu}_e + p \rightarrow \bar{e} + n$$
$$\nu_\mu + n \rightarrow \mu + p$$
$$\bar{\nu}_\mu + p \rightarrow \bar{\mu} + n$$

The detector works by detecting the light from the charged $e, \bar{e}, \mu, \bar{\mu}$ particles which is emitted when these charged particles interact with the electrons in the water atoms as they travel through the water. Depending on their energy, these particles can travel a long way – from several metres to hundreds of metres before stopping. This light is collected by phototubes on the walls of the water tank, and from its properties, the energy and information on the identity of the particle can be gathered. With the experimental setup used, it is possible to distinguish electrons and muons, but not electrons from positrons or muons from anti-muons.

While there are significant uncertainties in the estimated number of atmospheric neutrinos, the proportion of $\nu_e, \bar{\nu}_e$ to $\nu_\mu, \bar{\nu}_\mu$ is known quite precisely – to within about 5%. The number of muon neutrinos should be about twice the number of electron neutrinos which can be figured out just from the known way in which muons, pions and kaons decay (which has been extremely well studied in the laboratory). This means that they should produce twice as many muons as electrons in their detector. But this is not what was found in the experiments. It was found that this ratio was only about 1.2 instead of 2.

When something unexpected is found, people usually study it in more detail, however the anomaly didn't go away. This result appeared in several different experiments, with the first significant results published in 1988, and in more detail in 1992. At that time it was proposed that neutrino oscillations may be occurring for the atmospheric neutrinos. After all, very large distances were being travelled by the neutrinos. Neutrinos produced on the other side of the Earth were travelling all the way through the Earth before reaching the detector. That is, a distance of about 10,000 km.

Meanwhile, no evidence for proton decay was obtained. Over time, the 'background' signal from atmospheric neutrinos became more and more interesting than the expected signal from proton decay which didn't emerge. Today, there is still no evidence for proton decay which means that the average lifetime of the proton must be greater than about 10^{33} years (for the GUT decay mode, $p \rightarrow \bar{e} + \pi^0$). However, as a consolation prize Georgi and collaborators were awarded the 2000 Dirac medal for predicting proton decay which eventually led to the important discovery of neutrino oscillations of the muon neutrino!

The impact of super-Kamiokande

A great experimental improvement came in April 1996 when a large new detector called super-Kamiokande began operations (see **Figure 8.1**). The main purposes of the super-Kamiokande experiment is to check and clarify the origin of the atmospheric neutrino anomaly, study neutrinos from the Sun and to continue the search for proton decay. It would also give very useful information if another nearby supernova exploded. The detector is located about a kilometre underground in a mine a few hundred kilometres from Tokyo. Super-Kamiokande is very similar to its predecessor, the Kamiokande detector, but about 20 times larger. It contains approximately 50,000 tons of pure water.

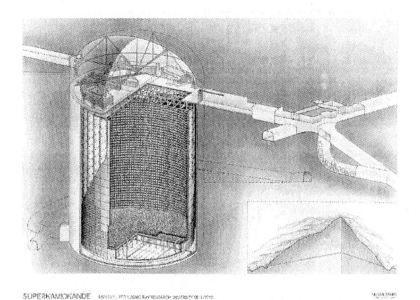

SUPERKAMIOKANDE INSTITUTE FOR COSMIC RAY RESEARCH UNIVERSITY OF TOKYO

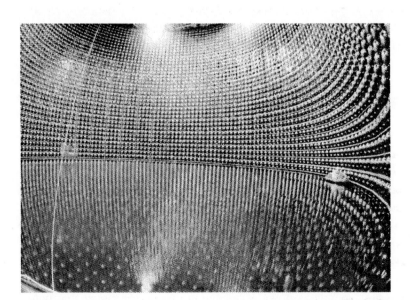

Figure 8.1a,b: The super-Kamiokande detector. From ICRR (Institute of cosmic ray research), The University of Tokyo.

Because of its huge size, the super-Kamiokande detector was able to quickly collect a very large sample of data. Not only did super-Kamiokande confirm the previously measured anomalously small value for the proportion of muons to electrons coming from atmospheric ν_μ and ν_e, but they were able to do something really new. Super-Kamiokande collected a very large sample of high energy events, that is, events with energy greater than a GeV. [Previous experiments were not large enough to collect a significant number of high energy events].

These high energy events are particularly useful because the detector does not 'see' the neutrinos themselves, but only the electrons and muons that are produced from the neutrino interactions. The direction of the electrons and muons is correlated with the direction of the neutrino, but at lower energies, where most of the data resides, this correlation is rather poor. The estimated average angle between the unobserved neutrino and the detected electron or muon is about 60° for the events with energies less than about a GeV. However, for the events with energies greater than about a GeV, the correlation is much better – about 15°. With the large amount of data that super-Kamiokande collected and is still collecting, the directional dependence of the anomaly could be studied in detail. They found that there was about half as many up-going muons as downward going muons produced from atmospheric neutrino interactions in their detector. For electrons they found the same number of up-going events as down-going events. **Figure 8.2** illustrates the basic geometry of the super-Kamiokande atmospheric neutrino experiment.

From simple symmetry arguments the expected number of up-going and down-going events should be the same (if the neutrinos are massless and therefore do not oscillate). The fact that for the up-going events the neutrinos traverse the whole of the Earth is of no consequence. The probability that they would have interacted while travelling this distance is very tiny. The interaction of neutrinos with matter is very well understood from countless laboratory experiments. Thus, the fact that they have found half as many up-going muons as down-going muons is a clear signal that new neutrino physics must exist. Could neutrino oscillations explain the missing muons?

Down-going neutrinos

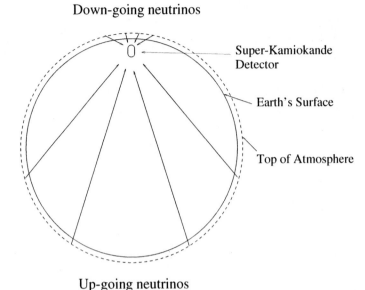

Figure 8.2: Geometry of the super-Kamiokande atmospheric neutrino experiment. The down-going neutrinos travel over very short distances (~ 20 km) and typically do not have time to oscillate, while the up-going neutrinos travel large distances ($\sim 10{,}000$ km), and have time for several oscillations.

Clearly, the fact that the number of up-going muons was less than expected immediately suggests muon neutrino (ν_μ) oscillations. Down-going events originate from neutrinos which typically travel very short distances $\lesssim 100$ kilometres, while up-going events originate from neutrinos which travel relatively large distances $\gtrsim 3000$ kilometres. Clearly, if $100\text{ km} \lesssim L_{osc} \lesssim 3000\text{ km}$, then there would be almost no effect for the down-going neutrinos because they don't have time to oscillate. On the other hand, in the case of up-going neutrinos, they would typically have time for several oscillations. This means that, on average, about half of the up-going muon neutrinos would have oscillated (for maximal mixing) into something else by the time they reached the detector. If there were only half as many up-going muon neutrinos (ν_μ) then the number of muons produced in the detector would be correspondingly reduced, thereby explaining the observed muon deficit. But what does the muon neutrino oscillate into?

The atmospheric neutrino anomaly and me

During 1993 I first became aware of the atmospheric neutrino anomaly. Even though the atmospheric neutrino anomaly had been around for some years before 1993, I was unaware of it. It wasn't emphasised much in the literature and it wasn't talked about much in conferences. Nevertheless, when I did finally hear about the atmospheric neutrino anomaly, it immediately occurred to me that it could be simply explained with the mirror matter theory.

The idea is very simple. As I have discussed in the previous chapter, the ordinary neutrino-mirror neutrino transition force or 'mass mixing' force leads to maximal oscillations of each of the known neutrinos with their mirror partners. It therefore seems very natural to explain the atmospheric neutrino anomaly via $\nu_\mu \to \nu'_\mu$ oscillations (where ν'_μ is the mirror muon neutrino). With the new results from the super-Kamiokande experiment the prediction of maximal mixing has been confirmed. This experiment can only be explained if the muon neutrino oscillates with a mixing angle within the range:

$$0.85 \stackrel{<}{\sim} \sin^2 2\theta \leq 1.0$$

This is in nice agreement with the 1993 mirror matter prediction of $\sin^2 2\theta = 1$.

Neutrino oscillations of the muon neutrino into its mirror partner is not the only possible solution to the atmospheric neutrino anomaly. A more conservative solution, not involving the mirror world, is possible if muon neutrinos oscillate with tauon neutrinos with near maximal mixing. Moreover, the super-Kamiokande data currently slightly prefers $\nu_\mu \to \nu_\tau$ oscillations, but the data is not sensitive enough to be able to convincingly exclude the possibility of $\nu_\mu \to \nu'_\mu$ oscillations. Nevertheless, the simplest mirror world solution to the atmospheric neutrino anomaly does at least provide an elegant explanation for the inferred maximal muon neutrino oscillations, which is a good reason to take it seriously. [Another good reason is that it is an explanation which is consistent with the LSND experiment as I will explain in a moment]. Fortunately several new experiments currently planned to start in 2006 should help clarify things. There is one European experiment and one US experiment. Both experiments will send a beam of ν_μ's over a distance of about 730 kilometres. There is nothing particularly special about the number 730, just one of those strange coincidences. In the European experiment the neutrinos are produced at the CERN Laboratory near Geneva in Switzerland, and will be detected again in the Gran Sasso Laboratory near Rome in Italy. Unfortunately, this experiment, like Rome, cannot be built in a day but will take five years or so. In the US experiment, the neutrino source is in Fermi National Laboratory near Chicago, and will be detected in the Soudan Laboratory, which is not in Africa but is in Minnesota.

The idea of these long baseline experiments is that the disappearance of ν_μ's from the atmospheric neutrino measurements can be checked, thereby making certain that maximal ν_μ oscillations are actually occurring. These experiments can also do something else. They can sensitively distinguish between the $\nu_\mu \to \nu'_\mu$ and $\nu_\mu \to \nu_\tau$ possibilities. They can do this by searching for τ particles which can only be produced from ν_τ interactions. [The τ particle has properties similar to an electron except that it is about 3000 times heavier, and decays rapidly]. Hence if the atmospheric neutrino anomaly is due to $\nu_\mu \to \nu'_\mu$ oscillations then no τ particles will be produced, while if the anomaly is due to $\nu_\mu \to \nu_\tau$ oscillations, lots of τ particles should be found.

LSND experiment

Evidence for neutrino oscillations has also emerged from yet another type of experiment. The Liquid Scintillator Neutrino Detector (LSND) experiment in Los Alamos has been searching for $\bar{\nu}_\mu \to \bar{\nu}_e$ oscillations in several ways. In this experiment a beam of $\bar{\nu}_\mu$

Figure 8.3: The LSND detector. Picture from the LSND Collaboration website (http://www.neutrino.lanl.gov/LSND/).

neutrinos is obtained from the decays of $\bar{\mu}$ particles. The idea is to search for $\bar{\nu}_e$ some distance away by the interaction process:

$$\bar{\nu}_e + p \rightarrow \bar{e} + n.$$

The neutron is subsequently captured releasing a gamma ray. The detector works by looking for a gamma ray correlated with the positron, $\bar{e}$. Thus, because they start with an almost pure $\bar{\nu}_\mu$ beam they can sensitively search for $\bar{\nu}_\mu \rightarrow \bar{\nu}_e$ oscillations[*] by looking for the appearance of $\bar{\nu}_e$. With their experimental setup the $\bar{\nu}_\mu$ source is located about 30 metres from the centre of their detector (the LSND detector is shown in **Figure 8.3** above).

---

[*]There are simple theoretical reasons suggesting that the rate of particle and anti-particle oscillations should be equal. Thus, observation of $\bar{\nu}_\mu \rightarrow \bar{\nu}_e$ oscillations implies that $\nu_\mu \rightarrow \nu_e$ oscillations must also exist at the same rate.

The data from the experiment suggests that $\nu_\mu \to \nu_e$ oscillations do indeed occur, however only a very small effect is observed. That is, the oscillation probability was found to be non-zero but small. From chapter 7 we learned that the oscillation probability has the form:

$$P(\nu_e \to \nu_\mu) = \sin^2 2\theta \sin^2 (L/L_{osc}).$$

The quantity L_{osc} is the oscillation length which depends on the neutrino energy and the neutrino masses, while L is the distance travelled by the neutrinos in the experiment. In the case of the LSND experiment, this distance was about 30 metres. Anyway, the small oscillation probability indicated by the experiment means that either the oscillation length is much longer than the distance travelled by neutrinos in the experiment (that is, $L_{osc} \gg L = 30$ metres), and/or the mixing angle $\sin^2 2\theta$ is very small. It turns out that some other experiments are able to exclude the possibility of a large oscillation length so that the only consistent explanation for this experiment is that *small angle* $\nu_\mu \to \nu_e$ oscillations occur, that is, $\sin^2 2\theta$ is close to zero.

Of course, it is only one experiment and another more sensitive experiment needs to be done to check it. It's always best to double-check things just to make sure. That's one good thing about experiments. They can always be improved and repeated so that all doubts can be removed, although this process can take time. To achieve this, a new experiment called BooNE is currently being constructed at Fermi National Laboratory near Chicago, and it is scheduled to begin data taking early in 2002. Within one year of operation it should be able to either convincingly confirm or refute the LSND claims.

If we accept for now that small angle $\nu_\mu \to \nu_e$ oscillations actually occur, as alleged by the LSND people, then how does this fit in with mirror matter? Actually it fits in very well because in the mirror matter theory the solar and atmospheric neutrino deficits appear to be most naturally explained by $\nu_e \to \nu_e'$ and $\nu_\mu \to \nu_\mu'$ oscillations respectively. This leaves open the possibility of oscillations between the families such as $\nu_\mu \to \nu_e$. The fact that the LSND experiment suggests that the oscillations involve a small angle fits in nicely with the general theoretical prejudice that mixing between families should be small. Of course, one must always be suspicious of theoretical prejudices since sometimes – perhaps even often – they turn out to be wrong.

Discussion

Neutrino physics represents a fascinating way in which the mirror matter theory may impact on particle physics experiments. The characteristic 'maximal mixing' prediction of the theory is apparently seen for both ν_e and ν_μ neutrino types. Furthermore 'small angle' oscillations have also allegedly been observed between the ν_e and ν_μ neutrino types, which can be accommodated within the mirror matter theory. In other words, the oscillation pattern suggested by the theory and the data is the following one:

Further experiments can really test whether these oscillations are occurring with the properties consistent with this theory.

I would like to emphasise that the above solution is the *not* the only possibility, although it does seem to be the most elegant one if mirror particles exist. The main alternative scenarios are:

- The solar neutrino problem is due to approximately maximal $\nu_e \to \nu_\tau$ oscillations, while the atmospheric neutrino anomaly is solved by $\nu_\mu \to \nu'_\mu$ oscillations.

- The solar neutrino problem is explained by maximal $\nu_e \to \nu'_e$ oscillations, while the atmospheric neutrino anomaly is due to $\nu_\mu \to \nu_\tau$ oscillations.

- The solar neutrino problem is due to maximal oscillations of $\nu_e \to (\nu_\mu + \nu_\tau)/\sqrt{2}$, while the atmospheric neutrino anomaly is due to maximal $\nu_\mu \to \nu_\tau$ oscillations. This scenario is called 'bi-maximal' neutrino oscillations and is currently quite popular among particle physicists.

Only the first two of these 'alternative' possibilities can accommodate the LSND signal for $\nu_e \to \nu_\mu$ oscillations. This means that the LSND experiment is really very important because if it is confirmed by BooNE, it suggests that either (or both) the atmospheric or solar neutrino deficits are solved by maximal oscillations into an invisible partner. Of course, the mirror matter theory not only provides a reason for an invisible (mirror) partner to exist, but also supplies the required maximal mixing too.

I summarize the glimpse of the mirror world which is indicated by the solar, atmospheric and LSND experiments in **Figure 8.4**. This concludes Part IV.

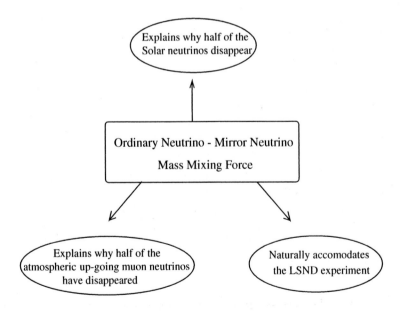

Figure 8.4: Neutrino wonders of the mirror world...

Chapter 9

Reflections on the Mirror World

One thing which has become increasingly clear over the last century of scientific endeavour is the importance of symmetries. We know that the fundamental interactions of nature are, to within experimental precision, completely symmetrical with respect to rotations of space, rotations of space-time and translations of space and time. However, the most obvious symmetry of all – left-right or mirror symmetry – is inexplicably not a symmetry of the known particles since the ordinary particles interact in a manifestly left-handed way. This experimental fact motivates the idea that a set of 'mirror particles' exist. The left-handedness of the ordinary particles can then be balanced by the right-handedness of the mirror particles so that it becomes possible for left-right symmetry to exist along side the other known symmetries.

It seems therefore, that the question I have been asking is a simple one – is mirror symmetry a fundamental symmetry of nature or not? Clearly, to answer this question we must look into the mirror. We must examine the observational and experimental consequences of mirror matter. If one bothers to look then it seems that there is an amazingly large amount of evidence for mirror matter and hence mirror symmetry:

- It predicts the existence of mirror matter in the Universe. Mirror matter would be invisible, making its presence felt by its gravitational effects. Remarkably there is a large body of evidence for such invisible 'dark' matter. There is also specific evidence that mirror stars have been observed from their gravitational effects on the bending of light.

- If mirror matter exists then mirror planets should also exist. In fact, there is remarkable evidence that these planets have actually been detected orbiting around nearby ordinary stars.

- There is evidence that the opposite type of system, with an ordinary planet orbiting a mirror star, also apparently exists, but has been misidentified as an 'isolated planet'!

- Perhaps most remarkable of all, is the evidence that mirror matter not only exists in our solar system, but mirror matter asteroid or comet sized objects are frequently colliding with our planet. There should even be fragments of mirror matter at various impact sites around the world, which could potentially be found. Nobody has looked!

On the microscopic level two types of forces or interactions can connect ordinary and mirror matter. That is, by small transition forces connecting photons with mirror photons and by small mass mixing terms between neutrinos and mirror neutrinos. The observational consequences of these effects are actually observed:

- The photon-mirror photon transition force implies a shorter effective lifetime for orthopositronium (a type of atom made from an electron and a positron) in vacuum experiments. A shorter lifetime is seen!

- Neutrino-mirror neutrino mass mixing implies that each ordinary neutrino oscillates into its mirror neutrino partner maximally. This means that over large distances precisely half of the neutrinos should be transformed into invisible mirror neutrinos. Again this is just what's observed! Half of the electron neutrinos emitted by the Sun are missing.

- There are three different types of neutrinos not one. In addition to electron neutrinos which are emitted from the Sun there are muon and tauon neutrinos too. Each of these neutrinos should also oscillate into their mirror partner if they travel over a large distance. While there are no experimental tests for tauon neutrino oscillations over a large distance, cosmic ray interactions in the atmosphere produce muon neutrinos. Remarkably, precisely half of these neutrinos have also apparently disappeared – again the expected result if mirror matter exists.

Do these seven pieces of evidence constitute seven pieces of a single jig-saw puzzle? Do they really constitute seven wonders of the mirror world implying that mirror matter exists? It is certainly conceivable that each of these puzzles could have other explanations. It is therefore still possible that mirror matter might not exist. I myself don't believe it absolutely – but I am curious, very curious to know if it really does exist.

To make certain that mirror matter is real we must repeat the orthopositronium experiment. We must clarify the neutrino anomalies and we must keep studying dark matter. Last, but not least, we must keep thinking... I am optimistic though that if the majority of these seven puzzles are solved by mirror matter then we will know within five years. In the meantime there is good reasons to be cautiously optimistic. Indeed, it is striking that every possible implication of mirror matter has actually been observed. Einstein once said[65]: 'The Lord is subtle but not malicious'. If mirror matter does not exist, why would all of its predictions have come true? This would be somewhat malicious... Of course, if mirror matter is nature's truth then it would be another important step in understanding nature's grand design. I hope that it won't be the last.

Further Reading

There are over 50 published scientific papers discussing various aspects of the mirror matter theory. In the following bibliography the most relevant papers pertaining to the issues discussed in this book are listed. These papers (and references there-in) might be useful to people interested in following up things in more detail.

Chapter 1 & 2

The idea of mirror matter first appeared in: "Question of Parity conservation in weak interactions", by T. D. Lee and C. N. Yang, *Physical Review*, Vol. 104, (1956), p. 254. From the 1960s - 1980s the idea was also discussed in several publications: "On the possibility of experimental observation of mirror particles", by I. Kobzarev, L. Okun and I. Pomeranchuk, *Soviet Journal of Nuclear Physics*, Vol. 3, (1966), p. 837; "External Inversion, Internal Inversion, and Reflection Invariance", by M. Pavsic, *International Journal of Theoretical Physics*, Vol. 9, (1974), p. 229; "On possible effects of mirror particles", by S. Blinnikov and M. Khlopov, *Soviet Journal of Nuclear Physics*, Vol. 36, (1982), p. 472. The idea was rediscovered and put into a modern context in the paper: "A model of fundamental improper space-time symmetries", by R. Foot, H. Lew and R. R. Volkas, *Physics Letters*, Vol. B272, (1991), p. 67.

Chapter 3

The general idea that mirror matter might be the dark matter in the Universe appears to have been first discussed in the paper by S. Blinnikov and M. Khlopov listed above. That mirror stars should appear in the gravitational microlensing experiments was predicted in: "Mirror baryons as the dark matter", by H. M. Hodges, *Physical Review*, Vol. D47, (1993), p. 456. It was further discussed in:

"Neutrino mass and the mirror universe", by Z. Silagadze, *Physics of atoms and the nucleus*, Vol. 60, (1997), p. 272 (hep-ph/9503481)[*]; "A quest for weak objects and for invisible stars", by S. Blinnikov, astro-ph/9801015 (1998); "Have mirror stars been observed?", by R. Foot, *Physics Letters*, Vol. B452, (1999), p. 83 (astro-ph/9902065).

Chapter 4

The idea that the close-in extrasolar planets might be mirror worlds was first discussed in: "Have mirror planets been observed?", by R. Foot, *Physics Letters*, Vol. B471, (2000), p. 191 (astro-ph/9908276). That they are opaque: "Are mirror worlds opaque?", by R. Foot, *Physics Letters*, Vol. B505, (2001), p. 1 (astro-ph/0101055). The idea that isolated planets are ordinary planets orbiting mirror stars was put forward in: "Do 'isolated' planetary mass objects orbit invisible mirror stars?", by R. Foot, A. Yu. Ignatiev and R. R. Volkas, Astroparticle Physics, in press (astro-ph/0010502).

For the latest observational/experimental situation on the extrasolar planets, see the 'extrasolar planet encyclopaedia' available at: http://cfa-www.harvard.edu/planets/encycl.html

Chapter 5

The implications of photon-mirror photon transition interactions for orthopositronium was discussed in: "Positronium versus the mirror Universe", by S. L. Glashow, *Physics Letters*, Vol. B167, (1986), p. 35. The importance of orthopositronium collisions on negating the effects of oscillations was discussed in "Limit on 'disappearance' of orthopositronium in vacuum", by S. N. Gninenko, *Physics Letters*, Vol. B326, (1994), p. 317. The detailed fit to the experiments showing that the experimental results could be explained by photon-mirror photon transitions was done in: "Can the mirror world explain the orthopositronium lifetime anomaly?", by R. Foot and S.

---

[*]Papers with astro-ph or hep-ph numbers can be downloaded directly from the web at the Los Alamos Preprint arXiv. For example for this paper with number hep-ph/9503481 go to the site: http://xxx.lanl.gov/format/hep-ph/9503481 (for other papers just change the number at the end and/or change hep-ph to astro-ph). The papers are available in postscript or pdf formats.

N. Gninenko, *Physics Letters*, Vol. B480, (2000), p. 171 (hep-ph/0003278). The astrophysical implications of the photon-mirror photon transition force was discussed in: "Physics of Mirror Photons", by R. Foot, A. Yu. Ignatiev and R. R. Volkas, *Physics Letters*, Vol. B503, (2001), p. 355 (astro-ph/0011156). The idea that Gamma Ray Bursts might be exploding mirror stars was proposed in: "Gamma Ray Bursts produced by mirror stars", by S. Blinnikov, astro-ph/9902305 (1999).

Chapter 6

Implications of mirror matter for Tunguska, Moon etc.: "Seven and a half reasons to believe in mirror matter: From neutrino puzzles to the inferred dark matter in the Universe", by R. Foot, *Acta Physica Polonica*, Vol. B32, (2001), p. 2253 (astro-ph/0102294) and "The mirror world interpretation of the 1908 Tunguska event and other more recent events", by R. Foot, *Acta Physica Polonica*, Vol. B32, (2001), p. 3133 (hep-ph/0107132). Implications of mirror matter for Nemesis: "TeV scale gravity, Mirror Universe and ....Dinosaurs", by Z. Silagadze, *Acta Physica Polonica*, Vol. B32, (2001), p. 99 (hep-ph/0002255). For Murray's planet see also: "Do mirror planets exist in our solar system?", by R. Foot and Z. Silagadze, *Acta Physica Polonica*, Vol. B32, (2001), p. 2271 (astro-ph/0104251). Finally, the mirror world explanation for the pioneer spacecraft anomalies was discussed in: "A mirror world explanation for the pioneer spacecraft anomalies?", by R. Foot and R. R. Volkas, *Physics Letters*, Vol. B517 (2001), p. 13 (hep-ph/0108051).

Chapter 7 & 8

The idea that neutrinos could oscillate into mirror neutrinos and thereby solve the solar neutrino problem was first discussed in: "Possible consequences of parity conservation", by R. Foot, H. Lew and R. R. Volkas, *Modern Physics Letters*, Vol. A7, (1992), p. 2567 and extended to the atmospheric neutrino anomaly in "Neutrino oscillations and the exact parity model", by R. Foot, *Modern Physics Letters*, Vol. A9, (1994), p. 169 (hep-ph/9402241) and further developed in "Neutrino Physics and the mirror world: How exact parity symmetry explains the solar neutrino deficit, the atmospheric neutrino anomaly and the LSND experiment", by R. Foot and R. R. Volkas, *Physical Review*, Vol. D52, (1995), p. 6595 (hep-ph/9505359).

The current experimental situation for each experiment can be obtained from the appropriate web pages:

SNO Homepage: http://www.sno.phy.queensu.ca

Borexino Homepage: http://almime.mi.infn.it

Super-Kamiokande Homepage:
http://www-sk.icrr.u-tokyo.ac.jp/doc/sk/

KamLAND Homepage:
http://www.awa.tohoku.ac.jp/html/KamLAND/index.html

LSND: http://www.neutrino.lanl.gov/LSND/

BooNE: http://www-boone.fnal.gov

For other atmospheric, solar and related neutrino experiments, see the 'Neutrino oscillation industry' website:

http://www.hep.anl.gov/ndk/hypertext/nuindustry.html

Notes

1. I. Newton, quoted in *The life of Isaac Newton*, by R. Westfall, Cambridge University Press, 1993, p. 309.

2. *A brief history of time*, by S. Hawking, Bantam Books, Toronto, 1988, p. 156.

3. W. Pauli, Letter to Schafroth, December 1956, quoted in *Symmetries in Physics (1600-1980)*, referred to as *SP* below, Proceedings of the 1st International Meeting on the History of Scientific Ideas, Edited by M. G. Doncel, A. Hermann, L. Michel, A. Pais, Servei de Publicacions, UAB, 1983, p. 352.

4. W. Pauli, Letter to Ray Davis, December 1956, quoted in *SP*, p. 352.

5. There is earlier discussion of the optical properties of the letters 'b' and 'd' by I. Asimov, *The left hand of the electron*, Panther, 1976, p. 35.

6. E. Rutherford, repr. in *Rutherford and the nature of the atom*, E. N. da C. Andrade, Doubleday, New York, 1964, p. 111.

7. P. Dirac, quoted in *Advanced Quantum Mechanics*, by J. J. Sakurai, Addison-Wesley, 1967, p. 1.

8. *One, Two, Three... Infinity*, by G. Gamow, Viking Press, New York, 1947, p. 69.

9. "Question of Parity conservation in weak interactions", by T. D. Lee and C. N. Yang, *Physical Review*, Vol. 104, 1956, p. 254.

10. J. Encke, Letter to Leverrier, quoted in *Planet Quest* by Ken Croswell, The Free Press, 1997, p. 44.

11. "Submillimetre tests of the gravitational inverse square law", by C. D. Hoyle *et al*, *Physical Review Letters*, Vol. 86, 2001, p. 1418.

12. "Basic constituents of the visible and invisible matter – a microscopic view of the Universe", by D. Roy, physics/0007025, 2000.

13. "Gravitational microlensing by the galactic halo", *Astrophysical Journal*, by B. Paczynski, Vol. 304, 1986, p. 1.

14. "Gravitational microlensing results from MACHO", by the MACHO Collaboration, astro-ph/9611059, 1996.

15. "The MACHO Project: Microlensing Results from 5.7 Years of LMC Observations", by the MACHO Collaboration, *Astrophysical Journal*, Vol. 542, 2000, p. 281.

16. "Dark Halos around Spirals; no role for collisionless Cold Dark Matter particles", by A. Borriello and P. Salucci, astro-ph/0106251, 2001;

"Dark Matter in the centre of galaxies and galaxy clusters: ruling out the cold dark matter scenario?", by E. D'Onghia *et al*, astro-ph/0107423, 2001.

17. From the preface of *The fifth essence: the search for dark matter in the universe*, L. Krauss, New York, Basic Books, 1989.

18. "A Jupiter-mass companion to a solar-type star", by M. Mayor and D. Queloz, *Nature*, Vol. 378, 1995, p. 355.

19. "First Results from the Anglo-Australian Planet Search – A Brown Dwarf Candidate and a 51 Peg-like Planet", by C. G. Tinney *et al*, astro-ph/0012204, 2000.

20. From an interview with G. Marcy, "Out of this world", *New Scientist*, Vol. 169, 13/01/2001, p. 44.

21. "Absence of a planetary signature in the spectra of 51 Pegasi", by D. Gray, *Nature*, Vol. 385, 1997, p. 795.

22. "Detection of Planetary Transits Across a Sun-like Star", by D. Charbonneau, T. Brown, D. Latham and M. Mayor, astro-ph/9911436, 1999.

23. "HST Time-Series Photometry of the Transiting Planet HD 209458", by T. Brown *et al*, astro-ph/0101336, 2001.

24. "A Transiting Peg-like Planet", by G. Henry, G. Marcy, P. Butler and S. Vogt, *Astrophysical Journal*, Vol. 529, 2000, p. L41.

25. "A Pair of Resonant Planets Orbiting GJ 876", by G. Marcy *et al*, *Astrophysical Journal*, Vol. 556, 2001, p. 296.

26. G. Marcy, quoted in *Planet Quest* by Ken Croswell, The Free Press, 1997, p. 198.

27. "Analysis of the Hipparcos measurements of Upsilon Andromedae - A mass estimate for one of its planetary companions", by T. Mazeh, S. Zucker, A. Dalla Torre, and F. van Leeuwen, *Astrophysical Journal*, Vol. 522, 1999, L149.

28. "Stability and Chaos in the Upsilon Andromedae Planetary System", by G. Laughlin and F. C. Adams, *Astrophysical Journal*, Vol. 526, 1999, p. 881.

29. "tau Boo b: Not so bright, but just as heavy", by A. Collier Cameron *et al*, astro-ph/0012186, 2000.

30. Quoted in the article, "Free Floating Planets", by D. Schneider, *American Scientist*, May-June 2000.

31. "Discovery of Young, Isolated Planetary Mass Objects in the σ Orionis Star Cluster", by M. Zapatero Osorio *et al*, *Science*, Vol. 290, 2000, p. 103.

32. "Search for Millicharged Particles at SLAC", by A. Prinz *et al*, *Physical Review Letters*, Vol. 81, 1998, p. 1175.

33. "Order α^2 corrections to the decay rate of orthopositronium", by G. S. Adkins, R. N. Fell and J. Sapirstein, *Physical Review Letters*, Vol. 84, 2000, p. 5086.

34. S. Glashow, in *Interactions: a journey through the mind of a particle physicist and the matter of this world*, Warner Books, New York, 1988, p. 10.

35. "Precision measurement of the orthopositronium decay rate using the vacuum technique", by J. S. Nico, D. W. Gidley, A. Rich and P. W. Zitzewitz, *Physical Review Letters*, Vol. 65, 1990, p. 1344.

"Precision measurement of the orthopositronium vacuum decay rate using the gas technique", by C. I. Westbrook *et al*, *Physical Review*, Vol. A40, 1989, p. 5489.

"New measurement of the orthopositronium decay rate", by S. Asai, S. Orito and N. Shinohara, *Physics Letters*, Vol. B357, 1995, p. 475.

36. "Direct detection of galactic halo dark matter", by B. Oppenheimer *et al*, astro-ph/0104293, 2001.

"Faint, moving objects in the Hubble Deep Field: components of the dark halo", by R. Ibata *et al*, *Astrophysical Journal*, Vol. 524, 1999, p.L1.

37. H. Lipkin, in *Physics Today*, July 2000, p. 15.

38. From the paper: "A novel left-right symmetric model", by R. Foot and H. Lew, hep-ph/9411390, 1994.

39. From the article, *"The sky has split apart!"* – *The cosmic mystery of the century*, by R. A. Gallant, available at http://www.galisteo.com/tunguska/docs/splitsky.html

40. F. Vologzhin, letter to Kulik, in *Giant Meteorites*, referred to as *GM* below, by E. Krinov, Pergamon Press, 1966, p. 133.

41. L. Kulik, quoted in *GM*, p. 190.

42. L. Kulik, from a report for the Presidum of the Academy of Sciences, quoted in *GM*, p. 199.

43. L. Kulik, from a booklet entitled 'Beyond the Tunguska Meteorite', quoted in *GM*, p. 196.

44. "Search for microremnants of the Tunguska Cosmic Body", by G. Longo, R. Serra, S. Cecchini and M. Galli, *Planetary and Space Science*, Vol. 42, 1994, p. 163.

45. "The Tunguska Meteorite problem today", by N. V. Vasilyev, http://www.galisteo.com/tunguska/docs/tmpt.html

46. M. Chown, Private Communication.

47. "Investigation of a bright flying object over northwest Spain, 1994 January 18", by J. A. Docobo *et al*, *Meteoritics & Planetary Science*, Vol. 33, 1998, p. 57.

   "Greenland superbolide event of 1997 December 9", by H. Pedersen et al, *Meteoritics & Planetary Science*, Vol. 36, 2001, p. 549.

48. Jordan Astronomical Report (by Mohammad Odeh) available on the web at http://www.jas.org.jo/mett.html

49. This possibility was suggested to me by Z. Ceplecha.

50. "Periodicity of extinctions in the geological past", by D. Raup and J. Sepkoski, *Proceedings of the National Academy of Sciences*, Vol. 81, 1984, p. 801.

51. "Are periodic mass extinctions driven by a solar companion?", by D. Whitmire and A. Jackson, *Nature*, Vol. 308, 1984, p. 713.

    "Extinction of species by periodic comet showers", by M. Davis, P. Hut and R. Muller, *Nature*, Vol. 308, 1984, p. 715.

52. Encyclopaedia Britannica, William Benton Publisher, 1968, Vol. 18, p. 71.

53. "The satellites of Neptune and the origin of Pluto", by R. Harrington and T. Van Fladern, *Icarus*, Vol. 39, 1979, p. 131.

54. "Arguments for the presence of a distant large undiscovered solar system planet", by J. Murray, *Monthly Notices of the Royal Astronomical Society*, Vol. 309, 1999, p. 31.

55. "Cometary evidence for a massive body on the outer Oort clouds", by J. Matese, P. Whitman and D. Whitmire, *Icarus*, Vol. 141, 1999, p. 354.

56. *Dark Matter missing planets and new comets*, by T. Van Flandern, North Atlantic Books, 1993, p. 200.

57. "Study of the anomalous acceleration of pioneer 10 and 11", by J. Anderson *et al*, gr-qc/0104064, 2001.

58. W. Pauli, letter to physicists' gathering at Tubingen, December 4, 1930, reprinted in *Inward Bound*, by A. Pais, Oxford University Press (1986), p. 315.

59. Some historical material was gathered from the article, *How the Sun Shines*, by J. N. Bahcall, astro-ph/0009259, 2000.

60. Lord Kelvin, "On the Age of the sun's heat", *Macmillan's Magazine*, March 5, 1862, p. 288 [Quoted by Bahcall in Ref.59].

61. *Crime and punishment*, by F. Dostoyevsky, translated by D. McDuff, Viking, 1991 (first published in 1866), p. 251.

62. "Neutrino astronomy and lepton charge", by V. Gribov and B. Pontecorvo, *Physics Letters*, Vol. B28, 1969, p. 493.

63. "Unity of all elementary particle forces", by H. Georgi and S. Glashow, *Physical Review Letters*, Vol. 32, 1974, p. 438.

64. S. Glashow, *Charm of Physics*, Simon & Schuster 1991, p. 168.

65. A. Einstein, quoted in *Subtle is the Lord: the science and life of Albert Einstein*, by A. Pais, New York, Clarendon Press 1982.